CHEMICAL KINETICS

By the same author

ATOMIC STRUCTURE AND VALENCY

CHEMICAL KINETICS

FOR GENERAL STUDENTS
OF CHEMISTRY

B. STEVENS
M.A., D.Phil

Professor of Chemistry
University of South Florida

LONDON

CHAPMAN AND HALL

© B. Stevens 1961, 1970
First published 1961
by Chapman and Hall Ltd
11 New Fetter Lane, London EC4
Second edition, as a Science Paperback, published 1970

Reprinted 1971 and 1976

ISBN 0 412 20740 0

Printed in Great Britain
by Whitstable Litho Ltd, Whitstable, Kent

Distributed in the U.S.A. by Halsted Press,
a Division of John Wiley & Sons, Inc, New York

PREFACE

THIS small book is written primarily for the degree student who is reading chemistry as a secondary subject. The honours student has several excellent treatises and numerous comprehensive monographs at his disposal but the general student has less time available to distinguish those aspects of fundamental importance from such a wealth of detailed information.

With this in mind, the writer has tried to keep this monograph as short as possible so that the reader may still remember something of the earlier chapters as he approaches the last, and so acquire a perspective of the topic as a whole; the first-year honours student may benefit in the same way. The text is concerned not with *how* reactions are investigated—no book is a substitute for the laboratory—but with *why* such experiments are conducted and *what* we may conclude from the results. The theoretical treatments are by no means rigorous but are sufficient, it is hoped, to provide an understanding of the experimental parameters in terms of the properties of the reacting molecules themselves.

Perhaps the size of the book will encourage the student to carry it with him during his kinetics course so that it can be consulted as the mind inquires; if the answers are too far away the questions may be forgotten and the curiosity which prompted them discouraged.

B. S.

Sheffield
July 1961

CONTENTS

CHEMICAL KINETICS

1

INTRODUCTION

A. Chemical Change

A CHEMICAL process is one in which the identity of molecules is changed; consequently it involves the rupture and formation of chemical bonds.

Chemical changes are responsible for the reproduction and growth of living things as well as for the conditions under which this is possible; for example, harmful ultra-violet radiation from the sun is absorbed by a layer of ozone produced photochemically from oxygen in the upper atmosphere; carbon dioxide, a product of respiration and decay, is believed to be responsible for maintaining the earth's surface at a temperature some hundreds of degrees above that of outer space by absorbing heat which would otherwise be radiated by the earth back into space; and the atmospheric oxygen essential to animal life is a product of photosynthetic reactions taking place in plants.

In the service of man chemical reactions produce work in internal combustion engines, heat from fuels of organic origin and digestible food from unpalatable raw materials; the brewing, tanning, plastics, oil and chemical industries show an annual profit from molecular rearrangements of which the products are in greater demand than the reactants; whilst the photochemical reduction of silver bromide to silver is the foundation of the film industry.

B. The Extent of Chemical Change

In principle every chemical reaction is reversible and may be written in the form

$$\text{reactants} \rightleftharpoons \text{products}$$

At equilibrium, the relative concentrations of reactants and products, denoted by square brackets, are determined by the equilibrium constant

$$K = \frac{[\text{products}]}{[\text{reactants}]}$$

which is related to the standard free energy change $\Delta G°$ of the reaction at temperature T by the thermodynamic expression

$$\log_e K = -\frac{\Delta G°}{RT}$$

where R is the gas constant. Therefore

(a) if the reaction is accompanied by a large decrease in free energy, it proceeds virtually to completion, i.e. K is large;

(b) if $\Delta G°$ is large and positive, K is very small and no discernible change will take place in the absence of external radiation;

(c) the equilibrium concentrations of reactants and products are of the same order of magnitude if $\Delta G°$ is small or zero.

Although the change in free energy dictates the *extent* to which a reaction will take place, *thermodynamic considerations do not tell us how or how fast* it will proceed; for example trinitrotoluene may be safely handled under certain conditions, but the large decrease in free energy accompanying its decomposition under others is often forcibly demonstrated.

C. Chemical Kinetics

Chemical kinetics is concerned with the rate and mechanism of chemical change. Each change or reaction can be represented by a stoichiometric or balanced equation of which the following are well-known examples

$$2KMnO_4 + 16HCl = 2KCl + 2MnCl_2 + 8H_2O + 5Cl_2$$
$$2NaOH + H_2SO_4 = Na_2SO_4 + 2H_2O$$

If the same reactants give different products under different conditions then more than one reaction is involved; for example, ethyl alcohol is dehydrated by passage over heated alumina but is oxidised to acetaldehyde by a finely divided copper catalyst

$$C_2H_5OH \begin{cases} Al_2O_3 \longrightarrow C_2H_4 + H_2O \\ 300°C \\ Cu \longrightarrow CH_3CHO + H_2 \end{cases}$$

These are different reactions.

Most reactions are complex and consist of two or more elementary processes which usually proceed at different rates. The overall decomposition of nitrogen pentoxide gas, for example, may be represented stoichiometrically by

$$2N_2O_5 = 4NO_2 + O_2$$

but is believed to take place in the three following stages

$$N_2O_5 \rightarrow NO_2 + NO_3 \qquad \text{fast}$$
$$NO_2 + NO_3 \rightarrow NO_2 + NO + O_2 \qquad \text{slow}$$
$$NO + NO_3 \rightarrow 2NO_2 \qquad \text{fast}$$

The slowest of these elementary processes determines the overall rate.

By measuring reaction rates under various conditions it is possible to determine the nature of the elementary processes involved and so establish the reaction mechanism. *The aim of chemical kineticists is to predict the rate of any reaction under a given set of conditions from a knowledge of the physical properties of the molecules concerned;* so far this has been achieved only in a very few simple cases.

D. Homogeneous and Heterogeneous Processes

It is convenient to classify a reaction as

(a) *homogeneous* if it takes place in a single phase, i.e. in the gas phase or in solution;

(b) *heterogeneous* if it proceeds at a phase boundary, e.g. at the walls of the reaction vessel.

There is nothing fundamentally different between these two types of process, which often take place at the same time; however, since reaction rates are measured in terms of concentration in the homogeneous medium and not at the phase boundary, the treatments of homogeneous and heterogeneous reaction rates are formally different.

2

THE RATE OF REACTION

A. Definition and Units

THE rate of a chemical reaction is expressed as the variation in concentration of either reactants or products with time; the units of rate are usually moles per litre per second (moles lit^{-1} sec^{-1}) for reactions in solution and moles per cubic centimetre per second (moles cm^{-3} sec^{-1}) for gaseous processes.

The rate of the reaction

$$A + B \longrightarrow C$$

is given by

$$-\frac{d[A]}{dt} = -\frac{d[B]}{dt} = \frac{d[C]}{dt} \text{ moles lit}^{-1} \text{ sec}^{-1} \qquad (2.1)$$

where $[A]$, $[B]$ and $[C]$ denote the concentrations of A, B and C in moles per litre and the time t is measured in seconds. Since reactants A and B disappear as the reaction proceeds, $d[A]/dt$ and $d[B]/dt$ are negative quantities and must be preceded by a negative sign to express a positive rate.

B. Measurement of Reaction Rate

It is usual to conduct a preliminary investigation to establish
(a) the nature and extent of possible side reactions;
(b) the position of equilibrium;
(c) that the reaction proceeds at a conveniently measurable rate.
The rate is then measured at constant temperature by one of the following methods which allow the concentration of either reactants or products to be followed with time.

1. *Volumetric or gravimetric analysis* of samples removed from a reacting solution at certain time intervals as in ester hydrolysis

$$CH_3COOC_2H_5 + H_2O \rightarrow CH_3COOH + C_2H_5OH$$

where the samples may be titrated against standard alkali to determine the concentration of acid produced; it is essential to dilute

4

each sample immediately it is removed to reduce further reaction during analysis.

In his classic investigation of the system

$$H_2 + I_2 \rightleftharpoons 2HI$$

Bodenstein (1897) allowed the reaction to proceed for a certain time in a heated sealed quartz bulb which was then broken under dilute aqueous alkali at room temperature, the iodide and iodine being determined by volumetric analysis and the hydrogen gas collected and measured. This procedure was repeated using different bulbs for different reaction times.

2. *The measurement of pressure change* in a gaseous system reacting at constant volume requires that the reaction leads to a change in the total number of molecules present as in the pyrolysis (thermal decomposition) of nitrogen pentoxide

$$2N_2O_5 \longrightarrow 4NO_2 + O_2$$

The final pressure of this system should theoretically be two and a half times the initial pressure, but it is actually less than this due to the reversible side reaction

$$2NO_2 \rightleftharpoons N_2O_4$$

which must be taken into account.

This is a convenient way of following the pyrolysis of organic compounds in the gas phase, e.g. of acetaldehyde:

$$CH_3CHO = CH_4 + CO$$

and azomethane

$$CH_3N_2CH_3 = C_2H_6 + N_2$$

since the number of molecules produced invariably exceeds the number consumed.

3. *The measurement of change in some physical property* related to reactant or product concentration or both. This may be any of the following:

(a) *Refractive index* which has been used to trace the polymerisation of styrene

(b) *Specific volume* which increases for example during the dissociation of diacetone alcohol

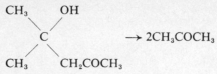

$$\longrightarrow 2CH_3COCH_3$$

but which usually decreases during polymerisation; in either case the reaction can be followed in a dilatometer as shown in Fig. 2.

(c) *Colour intensity* of reactant or product, e.g. a colorimeter may be used to measure the rate at which iodine is produced in the oxidation of hydriodic acid in solution

$$2HI + H_2O_2 = I_2 + 2H_2O$$

(d) *Specific rotation* of the plane of polarised light if the compounds are optically active. The first reliable quantitative investigation of the rate of a chemical reaction was that conducted by Wilhelmy (1850), who used a polarimeter to follow the inversion of dextrorotatory sucrose to a laevorotatory mixture of glucose and fructose in aqueous solution

$$\underset{\text{sucrose}}{C_{12}H_{22}O_{11}} + H_2O = \underset{\text{glucose}}{C_6H_{12}O_6} + \underset{\text{fructose}}{C_6H_{12}O_6}$$

(e) *Specific conductance* which decreases during ester saponification in aqueous solution

$$CH_3COOC_2H_5 + OH^- = CH_3COO^- + C_2H_5OH$$

since the mobility of the negative ion produced is much less than that of the hydroxyl ion consumed.

(f) *Viscosity* which increases during polymerisation.

(g) *Thermal conductivity:* the rate of decomposition of ammonia at the surface of an electrically heated tungsten wire is related to the increase in current required to maintain the wire at its initial temperature since the products of the heterogeneous reaction

$$2NH_3 = N_2 + 3H_2$$

conduct more heat from the wire to the walls of the reaction vessel.

4. *The measurement of heat liberated* in an exothermic reaction. The technique of differential thermal analysis involves the recording of temperature differences between calorimeters containing a reacting solution and pure solvent, respectively, as the temperature of both is increased at the same rate. This difference in temperature depends on the rate at which heat is liberated by the reaction which

in turn is related to the heat of the reaction and its rate. Thus from a single experiment it is possible to obtain

(i) the reaction rate at different temperatures;
(ii) the activation energy from (i);
(iii) the heat of reaction;
(iv) the activation energy of the reverse reaction from (ii) and (iii).

The decomposition of benzenediazonium chloride in aqueous solution

has been investigated in this way.

C. Analysis of Experimental Results

If the concentration of a reactant or product is plotted against the time elapsed since the beginning of the reaction, the reaction rate after a given time is numerically equal to the slope of the curve obtained at that time. Fig. 1 shows the rate of disappearance of hydrogen peroxide in alkaline solution at $40°C$ followed by titration.

$$2H_2O_2 = 2H_2O + O_2$$

The slope of the experimental curve is

$$\frac{d[H_2O_2]}{dt} = \begin{cases} \dfrac{0{\cdot}156 - 0{\cdot}060}{0 - 8400} = -1{\cdot}14 \times 10^{-5} \text{ moles lit}^{-1} \text{ sec}^{-1} \\ \qquad\qquad\qquad\qquad\qquad \text{at } t = 0 \\[2mm] \dfrac{0{\cdot}142 - 0{\cdot}060}{0 - 11\,300} = -0{\cdot}76 \times 10^{-5} \text{ moles lit}^{-1} \text{ sec}^{-1} \\ \qquad\qquad\qquad\qquad\qquad \text{after } 6000 \text{ sec} \end{cases}$$

The reaction rate under these conditions is therefore

$$-\frac{d[H_2O_2]}{dt} = \begin{cases} 1{\cdot}14 \times 10^{-5} \text{ moles lit}^{-1} \text{ sec}^{-1} \text{ initially} \\ 0{\cdot}76 \times 10^{-5} \text{ moles lit}^{-1} \text{ sec}^{-1} \text{ after} \\ \qquad\qquad\qquad 100 \text{ min} \end{cases}$$

Usually the measured quantity, which may be pressure, resistance, colour intensity, millilitres of standard alkali, etc., is plotted directly against time without conversion to concentration units. If the reaction is followed in a dilatometer for example, the measured quantity is the length L of liquid in the capillary which, in the case of dissociation of diacetone alcohol DA, increases with time as shown

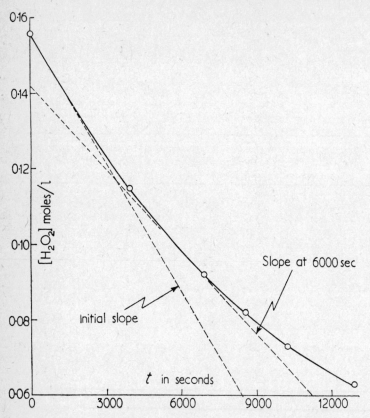

FIG. 1. Decomposition of H_2O_2 in alkaline solution at 40°C.
(*Burki and Schaaf, 1921*)

in Fig. 2 (Murphy, 1931). At any time t the reaction rate is related to the slope dL/dt of the curve at t by the expression

$$\text{rate} = -\frac{d[DA]}{dt} = -\frac{dL}{dt} \times \frac{d[DA]}{dL} \text{ moles lit}^{-1} \text{ sec}^{-1} \quad (2.2)$$

The difference between the initial and final capillary readings L_0 and L_∞ is proportional to the total amount of alcohol consumed in the reaction, i.e.

$$L_\infty - L_0 = CV([DA]_0 - [DA]_e) \text{ cm} \quad (2.3)$$

where C = a proportionality constant;
V = total volume of liquid (lit);
$[DA]_0$ = initial alcohol concentration (moles/lit);
$[DA]_e$ = final equilibrium concentration (moles/lit).

If L is the capillary reading and $[DA]$ the reactant concentration after t sec, then

$$L_\infty - L = CV([DA] - [DA]_e) \text{ cm} \quad (2.4)$$

whence from (2.3) and (2.4)

$$[DA] - [DA]_e = (L_\infty - L) \left\{ \frac{[DA]_0 - [DA]_e}{L_\infty - L_0} \right\} \text{ moles/lit}$$

or

$$\frac{d[DA]}{dL} = - \left\{ \frac{[DA]_0 - [DA]_e}{L_\infty - L_0} \right\} \text{ moles lit}^{-1} \text{ cm}^{-1} \quad (2.5)$$

and expression (2.2) becomes

$$\text{rate} = -\frac{d[DA]}{dt} = \frac{dL}{dt} \left\{ \frac{[DA]_0 - [DA]_e}{L_\infty - L_0} \right\} \text{ moles lit}^{-1} \text{ sec}^{-1} \quad (2.6)$$

From Fig. 2 we see that

$$\frac{dL}{dt} = \frac{22}{600} = 0.037 \text{ cm/sec}$$

when $t = 0$, $L_\infty = 28.7$ cm and $L_0 = 0.0$ cm.

Thus since $[DA]_0 = 0.17$ moles/lit and the reaction proceeds virtually to completion, i.e. $[DA]_e \approx 0$, the initial rate according to equation (2.6) is

$$0.037 \times \frac{0.17}{28.7} = 2.19 \times 10^{-4} \text{ moles lit}^{-1} \text{ sec}^{-1} \text{ at } 25°C$$

Expression (2.6) can be applied to any reaction followed by

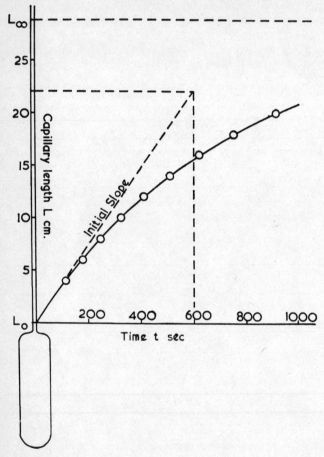

Fig. 2. Dissociation of diacetone alcohol followed in a dilatometer at 25°C. Alcohol concentration 0·17 moles/lit with K_2CO_3 catalyst.

measuring the change in some physical property X proportional to the concentration of reactant A, in which case it may be written

$$- \frac{d[A]}{dt} = \frac{dX}{dt} \left\{ \frac{[A]_0 - [A]_e}{X_\infty - X_0} \right\} \text{ moles lit}^{-1} \text{ sec}^{-1} \quad (2.7)$$

It is therefore necessary to determine

(a) the initial and final or equilibrium reactant concentrations $[A]_0$ and $[A]_e$;

(b) the initial and final values X_0 and X_∞ of the property measured.

If the reaction goes virtually to completion, $[A]_e$ may be neglected as in the dissociation of diacetone alcohol above, whilst if X is proportional to the concentration of a product D, equation (2.7) becomes

$$\text{rate} = \frac{d[D]}{dt} = \frac{dX}{dt} \times \frac{[D]_e}{X_\infty} \text{ moles lit}^{-1} \text{ sec}^{-1} \quad (2.8)$$

since $[D]_0 = 0$ and $X_0 = 0$.

Gaseous reactions are conveniently followed by recording the change in pressure observed if the total number of molecules changes. Fig. 3 shows a typical record of the change in pressure of the reacting system

$$CH_3CHO = CH_4 + CO$$

at 518°C. The rate after time t can be obtained from the slope dp/dt of the experimental curve using equation (2.7) which may be written

$$- \frac{d[CH_3CHO]}{dt} = \frac{dp}{dt} \left\{ \frac{[CH_3CHO]_0}{p_\infty - p_0} \right\} \text{ moles lit}^{-1} \text{ sec}^{-1} \quad (2.9)$$

if the aldehyde is completely decomposed. Since the concentration of a gas is $1/22.4$ moles/lit at 760 mm pressure and 273°K, then

$$[CH_3CHO]_0 = \frac{1}{22.4} \times \frac{p_0}{760} \times \frac{273}{(273 + 518)} \text{ moles/lit}$$

and expression (2.9) becomes

$$- \frac{d[CH_3CHO]}{dt} = \frac{dp}{dt} \times 2.03 \times 10^{-5} \text{ moles lit}^{-1} \text{ sec}^{-1} \text{ at } 518°C$$

since according to the stoichiometric equation

$$p_\infty = 2p_0$$

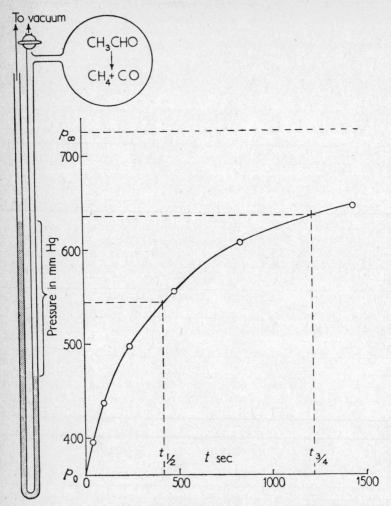

FIG. 3. Pyrolysis of acetaldehyde vapour at 518°C followed manometrically.

(*Hinshelwood, 1926*)

D. Factors Determining the Rate of Reaction

The rate of chemical change will depend primarily on the properties of the reacting molecules themselves; it is also found to vary with

(a) the concentrations of reacting molecules;

(b) the temperature;

so that *in order to understand why molecules react at the observed rate it is necessary to express this rate in terms of quantities which are independent of temperature and concentration.* Reaction rates are also affected by

(c) the presence of catalysts or inhibitors;

and may be influenced by

(d) the concentration of products;

(e) visible or ultra-violet light;

(f) ionising radiation, e.g. α-, β- and γ-rays.

The effects of these variables are discussed in turn in the following chapters.

3

THE DEPENDENCE OF REACTION RATE ON REACTANT CONCENTRATION

A. The Rate Expression

THE law of mass action, founded on experimental observation, may be written: *the rate of chemical change varies directly as the active concentrations of the reactants*. For the homogeneous reaction

$$A + B \longrightarrow \text{products}$$

in which the active concentrations of molecules A and B are equal to their actual concentrations in the reacting system, the rate is given by

$$- \frac{d[A]}{dt} = k[A][B] \text{ moles lit}^{-1} \text{ sec}^{-1} \tag{3.1}$$

This mathematical formulation of the mass action law is known as the *rate expression* in which the proportionality constant k is referred to as the *rate constant* for the particular reaction.

Since the reactant concentrations $[A]$ and $[B]$ decrease as the reaction takes place, the rate must also decrease with time according to (3.1). The rate constant k on the other hand remains unchanged throughout the reaction and provides a more convenient measure of reaction velocity than does the experimentally measured rate.

It should be noted that since the rate is proportional to the *active concentrations* of the reactants, then:

(a) The rate does *not* depend on the total *amount* of reactants present; e.g. 2 lit of a 1 M solution of benzenediazonium chloride will decompose at the same rate as 1 lit of the same solution at the same temperature, but 1 lit of a 2 M solution will decompose at twice the rate of either.

(b) For a homogeneous reaction between charged ions which are non-ideal solute particles the active concentrations are no longer equal to their stoichiometric concentrations and *ionic activities* must be used in the rate expression.

14

(c) The active concentrations of reactants in a heterogeneous process refer to the concentrations in the reaction zone; e.g. the rate at which ammonia decomposes on a heated tungsten wire is proportional to the number of NH_3 molecules *actually adsorbed* on the metallic surface.

B. Reaction Order

To obtain the characteristic rate constant k from experimental rate data it is necessary to know the number of concentration terms in the rate expression (3.1); the number of these terms, i.e. *the sum of the powers to which the reactant concentrations are raised in the rate expression is known as the reaction order.*

The experimental rate expression for the reaction

$$2NO + O_2 = 2NO_2$$

which takes place very quickly in the gas phase at room temperature and atmospheric pressure, is

$$-\frac{d[O_2]}{dt} = k[NO]^2[O_2] \text{ moles lit}^{-1} \text{ sec}^{-1}$$

i.e. the reaction is third-order, as might be expected from its stoichiometric equation. However, the reaction

$$3KClO = KClO_3 + 2KCl$$

follows the second-order rate expression

$$\frac{d[KClO_3]}{dt} = k[KClO]^2 \text{ moles lit}^{-1} \text{ sec}^{-1}$$

in solution whereas the stoichiometric equation predicts an order of three. This is simply due to the fact that the overall reaction represented by the stoichiometric equation is complex and is believed to take place in the stages

$$2KClO \longrightarrow KCl + KClO_2$$

$$KClO_2 + KClO \longrightarrow KCl + KClO_3$$

of which the first is rate-controlling.

It is therefore essential to establish the reaction order experimentally by one of the methods described below (Chapter 3D) since this

(a) allows the rate constant to be calculated from the measured rate;

(b) provides information concerning the mechanism of the overall reaction.

C. The Rate Constant

The rate constant k is numerically equal to the rate when the reactants are at unit concentration (3.1), but it is seldom possible to measure the rate under these conditions. It can be obtained from the initial rate if the initial reactant concentrations $[A]_0$ and $[B]_0$ are known, when (3.1) becomes

$$\text{initial rate} = k[A]_0[B]_0 \text{ moles lit}^{-1} \text{ sec}^{-1}$$

and

$$k = \frac{\text{initial rate}}{[A]_0[B]_0} \text{ lit mole}^{-1} \text{ sec}^{-1}$$

However, since it is often difficult to estimate the exact time at which the reaction starts it is preferable to integrate the rate expression (3.1) to obtain a relation between the rate constant and the amount of chemical change after an interval t; the particular form of integrated equation obtained depends on the number of concentration terms involved, i.e. on the reaction order.

First-order reactions

The rate expression

$$- \frac{d[A]}{dt} = k[A] \text{ moles lit}^{-1} \text{ sec}^{-1} \quad (3.2)$$

for the first-order process

$$A \longrightarrow \text{products}$$

can be rearranged to

$$- \frac{d[A]}{[A]} = -d \log_e [A] = k\,dt \quad (3.3)$$

If the initial concentration $[A]_0$ is reduced to $[A]_t$ after t sec then

$$- \int_{[A]_0}^{[A]_t} d \log_e [A] = k \int_0^t dt \quad (3.4)$$

Since

$$\int d \log_e [A] = \log_e [A] + \text{constant}$$

(3.4) becomes

$$-\log_e [A]_t + \log_e [A]_0 = \log_e \frac{[A]_0}{[A]_t} = kt$$

whence

$$k = \frac{1}{t} \log_e \frac{[A]_0}{[A]_t} \text{ sec}^{-1} \tag{3.5}$$

The units of k for first-order reactions can readily be verified from the rate expression (3.2), thus

$$\text{units of } k = \frac{\text{units of rate}}{\text{units of concentration}} = \frac{\text{moles lit}^{-1} \text{ sec}^{-1}}{\text{moles lit}^{-1}} = \text{sec}^{-1}$$

The first-order decomposition of benzenediazonium chloride in water

can be followed by measuring the volume V of nitrogen gas evolved as the reaction proceeds. Fig. 4 shows a typical volume–time record for this reaction at 50°C. Since

$$[C_6H_5N_2Cl]_0 \propto V_\infty$$

and

$$[C_6H_5N_2Cl] \propto (V_\infty - V_t)$$

equation (3.5) may be rearranged to

$$\log_e \left(\frac{V_\infty}{V_\infty - V_t} \right) = kt \tag{3.6}$$

$\text{Log}_{10}[V_\infty/(V_\infty - V_t)]$ is plotted as a function of t in Fig. 5 from the data in Fig. 4; the linear relationship confirms the order of the reaction and provides the value

$$k = 2 \cdot 303 \times \text{slope} = 6 \cdot 66 \times 10^{-2} \text{ min}^{-1}$$
$$= 1 \cdot 11 \times 10^{-3} \text{ sec}^{-1} \text{ at } 50°C$$

Theoretically a chemical reaction never stops since its rate decreases as the reactants are consumed and an infinite time is required for completion; however, the time required for the reactant concentration to be reduced to exactly one-half of its initial value is a characteristic interval known as the *time of half-change or reaction half-life* $t_{\frac{1}{2}}$. Since

$$[A]_t = [A]_0/2$$

when

$$t = t_{\frac{1}{2}}$$

expression (3.5) becomes

$$t_{\frac{1}{2}} = \frac{\log_e 2}{k} \text{ sec} \tag{3.7}$$

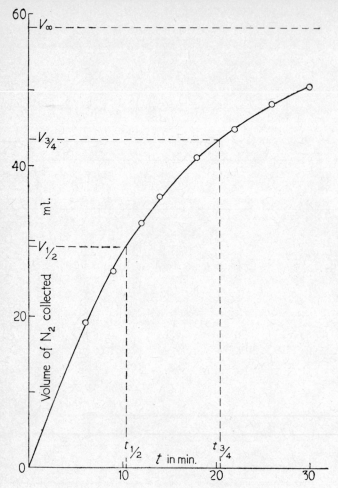

FIG. 4. Decomposition of benzenediazonium chloride at 50°C.
(*Crossley, Kienle and Benbrook, 1940*)

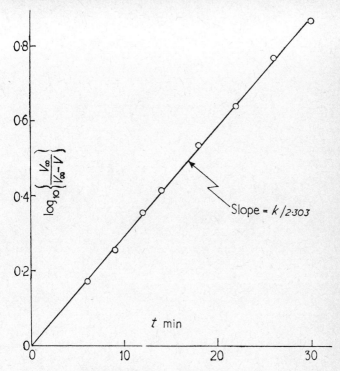

FIG. 5. Plot of data in Fig. 4 according to equation (3.6).

i.e. $t_{\frac{1}{2}}$ is independent of the initial reactant concentration for a first-order reaction. From Fig. 4 it is seen that $V = V_{\frac{1}{2}}$ after 625 sec; accordingly

$$k = \frac{\log_e 2}{625} = 1{\cdot}11 \times 10^{-3} \text{ sec}^{-1}$$

which is the value obtained from the slope in Fig. 5.

A classic example of a first-order process is that provided by radioactive decay where the characteristic half-lives vary from a millionth of a second for very unstable nuclei to millions of years for thorium and uranium. The pyrolyses of a number of organic vapours exhibit first-order behaviour under certain conditions although the overall decomposition is often complex and involves radical chains

(Chapter 10). Other examples of first-order reactions are given in Table 3.1.

TABLE 3.1. First-Order Reactions

In the gas phase

$$2N_2O_5 = 4NO_2 + O_2$$
$$CH_3N_2CH_3 = C_2H_6 + N_2$$
$$SO_2Cl_2 = SO_2 + Cl_2$$

In solution

$$\begin{matrix} CH_3 \\ \end{matrix}\!\!\!>\!\!C(OH)CH_2COCH_3 = 2CH_3COCH_3$$

$$2N_2O_5 = 4NO_2 + O_2$$

Second-order reactions

The rate expression for the second-order process

$$A + B \longrightarrow \text{products}$$

is

$$-\frac{d[A]}{dt} = k[A][B] \text{ moles lit}^{-1} \text{ sec}^{-1} \tag{3.8}$$

In the simplest case where $[A] = [B]$ this can be rearranged to

$$-\frac{d[A]}{[A]^2} = k\,dt \tag{3.9}$$

The integrated form of (3.9) between the limits $[A]_0$ at $t = 0$ and $[A]_t$ at $t = t$ sec is

$$-\int_{[A]_0}^{[A]_t} \frac{d[A]}{[A]^2} = k \int_0^t dt$$

or since

$$-\int \frac{d[A]}{[A]^2} = \frac{1}{[A]} + \text{constant}$$

$$\frac{1}{[A]_t} - \frac{1}{[A]_0} = \frac{[A]_0 - [A]_t}{[A]_0[A]_t} = kt$$

i.e. $$k = \frac{1}{t}\frac{[A]_0 - [A]_t}{[A]_0[A]_t} \text{ lit mole}^{-1} \text{ sec}^{-1} \qquad (3.10)$$

The dimensions of the second-order rate constant can readily be established from either (3.8) or (3.10).

At the time of half-change when

$$[A]_t = [A]_0/2$$

(3.10) becomes

$$t_\frac{1}{2} = \frac{1}{k[A]_0} \text{ sec} \qquad (3.11)$$

i.e. the half-life of a second-order process is inversely proportional to the initial reactant concentrations if these are equal.

Fig. 6 shows the hydroxyl concentration–time curve for the second-order reaction

$$CH_3COOC_2H_5 + OH^- = CH_3COO^- + C_2H_5OH$$

in aqueous ethanol at 30°C with

$$[CH_3COOC_2H_5]_0 = [OH^-]_0 = 0.05 \text{ moles/lit}$$

Under these conditions

$$t_\frac{1}{2} = 30 \text{ min} = 1800 \text{ sec}$$

Consequently

$$k = \frac{1}{t_\frac{1}{2}[OH^-]_0} = \frac{1}{1800 \times 0.05} = 0.011 \text{ lit mole}^{-1} \text{ sec}^{-1} \text{ at } 30°C$$

The rate constant can also be obtained from the integrated rate expression (3.10) as shown in Fig. 7, where

$$\frac{[OH^-]_0 - [OH^-]_t}{[OH^-]_0[OH^-]_t}$$

is plotted against t; the straight line obtained confirms that the reaction is second-order and provides the value

$$k = \text{slope} = 0.66 \text{ lit mole}^{-1} \text{ min}^{-1}$$
$$= 0.011 \text{ lit mole}^{-1} \text{ sec}^{-1}$$

which is that obtained from the half-life.

Other examples of second-order reactions are given in Table 3.2.

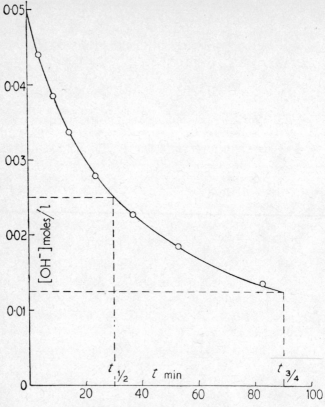

FIG. 6. Saponification of ethyl acetate at 30°C in aqueous alcohol
$[CH_3COOC_2H_5]_0 = [OH^-]_0 = 0.05$ moles/lit.
(*Smith and Levenson, 1939*)

Third-order reactions

The rate expression for the process:

$$A + 2B \longrightarrow \text{products}$$

is

$$-\frac{d[A]}{dt} = k[A][B]^2 \text{ moles lit}^{-1} \text{ sec}^{-1}$$

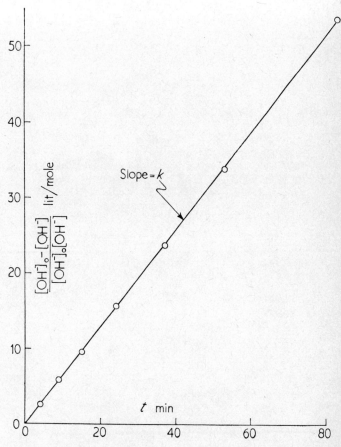

FIG. 7. Plot of data in Fig. 6 according to equation (3.10).

TABLE 3.2. Second-Order Reactions

In the gas phase

$$2HI = H_2 + I_2$$
$$H_2 + I_2 = 2HI$$
$$2NO_2 = 2NO + O_2$$
$$2NOCl = 2NO + Cl_2$$

In solution

$$3KClO = 2KCl + KClO_3$$
$$NH_4^+ + CNO^- = CO(NH_2)_2$$
$$C_2H_5ONa + CH_3I = C_2H_5OCH_3 + NaI$$
$$(C_2H_5)_3N + C_2H_5Br = (C_2H_5)_4N^+ + Br^-$$
$$CH_3COOH + C_2H_5OH = CH_3COOC_2H_5 + H_2O$$
$$CH_3COOC_2H_5 + OH^- = CH_3COO^- + C_2H_5OH$$
$$*CH_3COOC_2H_5 + H_2O = CH_3COOH + C_2H_5OH$$
$$*CH_3CONH_2 + H_2O = CH_3COOH + NH_3$$

* These are pseudo first-order in aqueous solution.

which rearranges to

$$- \frac{d[A]}{[A]^3} = k\,dt \tag{3.12}$$

if $[A] = [B]$. Integrating (3.12) between the limits $[A]_0$ at $t = 0$ and $[A]_t$ at $t = t$ sec leads to

$$- \int_{[A]_0}^{[A]_t} \frac{d[A]}{[A]^3} = k \int_0^t dt$$

or

$$\frac{1}{2[A]_t^2} - \frac{1}{2[A]_0^2} = kt$$

since

$$- \int \frac{d[A]}{[A]^3} = \frac{1}{2[A]^2} + \text{constant}$$

Thus

$$k = \frac{1}{2t} \left\{ \frac{1}{[A]_t^2} - \frac{1}{[A]_0^2} \right\} \text{ lit}^2 \text{ mole}^{-2} \text{ sec}^{-1} \tag{3.13}$$

The dimensions of third-order rate constants are again different from those of first- and second-order constants and can be deduced from the rate expression (3.12) in the same way.

Substitution of

$$[A]_t = [A]_0/2 \quad \text{when } t = t_{\frac{1}{2}}$$

in equation (3.13) leads to

$$t_{\frac{1}{2}} = \frac{3}{2k[A]_0^2} \text{ sec}$$

i.e. the half-life of a third-order process is inversely proportional to the square of the initial reactant concentration if these are equal.

Few third-order reactions are known since the rate-controlling process would be expected to involve a collision of all three reactant molecules. Some examples are given in Table 3.3.

TABLE 3.3. Third-Order Gas Reactions

$$2NO + O_2 = 2NO_2$$
$$2NO + Cl_2 = 2NOCl \qquad \text{All involve two molecules}$$
$$2NO + Br_2 = 2NOBr \qquad \text{of NO}$$
$$2NO + H_2 = N_2O + H_2O$$

A type of reaction which exhibits third-order kinetics is that in which atoms combine to give a diatomic molecule, e.g.

$$O + O + M \longrightarrow O_2 + M$$

where M is an atom or molecule which removes some of the vibrational energy of the newly formed O_2 molecule. Clearly if M did not fulfil this role the O_2 molecule would contain enough energy to redissociate immediately and no combination of atoms would take place. The frequency of collisions between three atoms or between two atoms and a molecule is proportional to the third power of concentration or gas pressure, consequently termolecular collisions are of little importance at the low pressures which prevail in the upper atmosphere where O_2 molecules are photochemically dissociated by light from the sun

$$O_2 + \text{light} \longrightarrow 2O$$

As the pressure and frequency of termolecular collisions decrease with altitude, the rate of recombination of O atoms becomes correspondingly slower until at an altitude of 60 miles or so the concentration of oxygen atoms is greater than that of oxygen molecules.

Any catalysed homogeneous second-order reaction is strictly third-order since the rate is proportional to the concentration of catalyst (Chapter 6); e.g. the rate expression for the acid-catalysed esterification

$$CH_3COOH + C_2H_5OH + H_3O^+$$
$$= CH_3COOC_2H_5 + H_2O + H_3O^+$$

is $\quad\dfrac{d[CH_3COOC_2H_5]}{dt}$

$$= k[CH_3COOH][C_2H_5OH][H_3O^+] \text{ moles lit}^{-1} \text{ sec}^{-1}$$

but since by definition the catalyst concentration $[H_3O^+]$ remains unchanged we may write

$$\frac{d[CH_3COOC_2H_5]}{dt} = k'[CH_3COOH][C_2H_5OH] \text{ moles lit}^{-1} \text{ sec}^{-1}$$

where $k' = k[H_3O^+]$ lit mol^{-1} sec^{-1}, so that this and similar reactions exhibit second-order kinetics.

General forms of the integrated rate expression

The simple forms of integrated rate expression (3.10) and (3.13) are valid only when the reactants are present in equal concentrations. If this is not the case integration of the appropriate expression is an exercise in calculus which is beyond the scope of this book, and the general forms shown in Table 3.4 should be used.

Pseudo-order reactions

If the concentration of one reactant is very much greater than that of the others, as for example when the solvent is one of the reactants in solution, the change in its concentration is negligible compared with the changes in concentration of the other molecules present and the overall order is reduced. Ester hydrolysis in aqueous solution

$$CH_3COOCH_3 + H_2O = CH_3COOH + CH_3OH$$

proceeds at the rate

$$\frac{d[CH_3COOH]}{dt} = k[CH_3COOCH_3][H_2O] \text{ moles lit}^{-1} \text{ sec}^{-1} \quad (3.14)$$

If the ester concentration is initially 1 mole/lit and is reduced to 0·5 moles/lit after t sec, i.e. its concentration changes by 50%, the concentration of water is reduced from

$$[H_2O] = 1000 \text{ g/lit} = 1000/18 \approx 55\cdot5 \text{ moles/lit}$$

to 55 moles/lit, i.e. by less than 1% in the same time. The concentration of water may therefore be regarded as constant and the rate expression (3.14) can be written

$$\frac{d[CH_3COOH]}{dt} = k'[CH_3COOCH_3] \text{ moles lit}^{-1} \text{ sec}^{-1}$$

where $\qquad\qquad k' = k[H_2O] \text{ sec}^{-1}$

TABLE 3.4. The General Form of Integrated Rate Expression

Rate expression	Integrated form
First-order $\quad -\dfrac{d[A]}{dt} = k[A]$	$kt = \log_e \dfrac{[A]_0}{[A]_t}$
Second-order $\begin{cases} -\dfrac{d[A]}{dt} = k[A]^2 \\[2mm] -\dfrac{d[A]}{dt} = k[A][B] \end{cases}$	$kt = \dfrac{[A]_0 - [A]_t}{[A]_0[A]_t}$ $kt = \dfrac{1}{[A]_0 - [B]_0} \log_e \left\{ \dfrac{[A]_t[B]_0}{[A]_0[B]_t} \right\}$
Third-order $\quad -\dfrac{d[A]}{dt} = k[A]^2[B]$	$kt = \dfrac{1}{(2[B]_0 - [A]_0)^2} \left\{ \dfrac{(2[B]_0 - [A]_0)([A]_0 - [A]_t)}{[A]_0[A]_t} + \log_e \dfrac{[A]_t[B]_0}{[A]_0[B]_t} \right\}$

The acid-catalysed inversion of sucrose in water

$$\text{sucrose} + H_2O + H_3O^+ \longrightarrow \text{glucose} + \text{fructose} + H_3O^+$$

behaves similarly, the experimental rate expression being

$$\frac{d[\text{glucose}]}{dt} = k''[\text{sucrose}] \text{ moles lit}^{-1} \text{ sec}^{-1}$$

with $\qquad\qquad k'' = k[H_2O][H_3O^+] \text{ sec}^{-1}$

Processes such as these which exhibit first-order kinetics although they involve more than one reactant are referred to as *pseudo first-order* reactions. It should be remembered that the pseudo-rate constants, k' and k'', contain concentration terms when attempts are made to account for their magnitudes on purely theoretical grounds (Chapter 5).

D. The Determination of Reaction Order

This is an essential part of the investigation of any chemical process and involves the analysis of concentration–time curves such as those shown in Figs. 1, 2, 3, 4 and 6.

Time of half-change

If the reactant concentrations are equal, the dependence of the reaction half-life $t_{\frac{1}{2}}$ on the initial concentration $[A]_0$ provides one of the most useful criteria of reaction order. It was shown above that

$$t_{\frac{1}{2}} = \begin{cases} \log_e 2/k \text{ sec for a first-order process} \\ 1/k[A]_0 \text{ sec for a second-order process} \\ 3/2k[A]_0^2 \text{ sec for a third-order process} \end{cases}$$

and in general for a reaction of overall order n

$$t_{\frac{1}{2}} \propto 1/[A]_0^{n-1} \text{ sec} \qquad (3.15)$$

or $\qquad\qquad \log t_{\frac{1}{2}} = \text{constant} - (n-1) \log [A]_0 \qquad (3.16)$

Hinshelwood and Green (1926) obtained the following half-lives for the homogeneous gas reaction

$$2NO + 2H_2 = N_2 + 2H_2O$$

at 826°C using equal initial reactant pressures p_0:

$p_0 =$	202	243	251	288	340·5	354	375 mm
$t_{\frac{1}{2}} =$	224	176	180	140	102	81	95 sec

These data are plotted according to (3.16) in Fig. 8 from which it is found that the overall order of the reaction is

$$n = 1 - \text{slope} = 1 - (-1\cdot85) = 2\cdot85$$

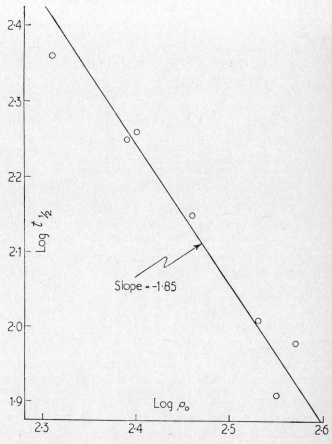

FIG. 8. Order of gas reaction $2NO + 2H_2 = N_2 + 2H_2O$ from equation (3.16).

which is more nearly third-order than any other; the small difference from the integral value is attributed to a simultaneous heterogeneous reaction at the walls of the quartz vessel.

The overall order may be obtained from a single concentration–time curve if the reaction is followed long enough for the time of three-quarters change $t_{\frac{3}{4}}$ to be estimated. After this interval $[A]_t = [A]_0/4$ whence from (3.5), (3.10) and (3.13)

$$t_{\frac{3}{4}} = \begin{cases} 2 \log_e 2/k = 2t_{\frac{1}{2}} & \text{for a first-order reaction} \\ 3/k[A]_0 = 3t_{\frac{1}{2}} & \text{for a second-order reaction} \\ 15/2k[A]_0{}^2 = 5t_{\frac{1}{2}} & \text{for a third-order reaction} \end{cases}$$

From the pressure–time curve shown in Fig. 3

$$t_{\frac{3}{4}} = 1220 \text{ sec and } t_{\frac{1}{2}} = 420 \text{ sec}$$

i.e. $t_{\frac{3}{4}} \approx 3t_{\frac{1}{2}}$ and the pyrolysis of acetaldehyde vapour is second-order under the conditions shown. The saponification of ethyl acetate is also second-order since from Fig. 6 we see that

$$t_{\frac{3}{4}} = 90 \text{ min and } t_{\frac{1}{2}} = 30 \text{ min}$$

whereas the decomposition of benzenediazonium chloride in aqueous solution (Fig. 4) follows first-order kinetics with

$$t_{\frac{3}{4}}/t_{\frac{1}{2}} = 20 \cdot 4/10 \cdot 2 = 2$$

The variation of $t_{\frac{1}{2}}$ with initial concentration and its relation to $t_{\frac{3}{4}}$ is a particularly important criterion of order in processes involving only one reactant; *if more than one reactant is involved this method requires that they are present in equal concentration*, which is not always experimentally possible.

The integration method

If k is calculated from the experimental concentration–time curves using the integrated rate expressions for first-, second- and third-order behaviour, the order which gives the most consistent values of k is the correct overall order. This is illustrated in Figs. 5 and 7, in which the experimental data in Figs. 4 and 6 are plotted on the assumptions that the reactions are first- and second-order respectively; the excellent linear relationship obtained in each case shows that these assumptions were correct. The data in Fig. 6 are plotted according to the first- and third-order expressions (3.5) and (3.13) in Fig. 9; clearly these do not represent the behaviour of this reaction.

It is again emphasised that the simple forms of the integrated rate expression (3.10) and (3.13) only apply if the initial reactant con-

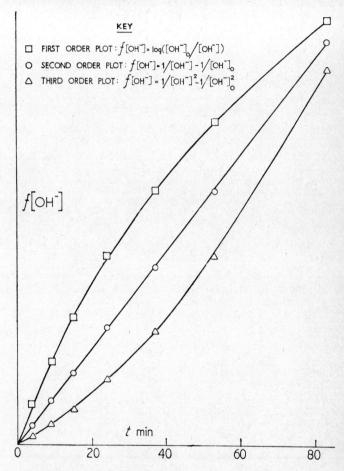

KEY

☐ FIRST ORDER PLOT: $f[OH^-] = \log([OH^-]_0/[OH^-])$

○ SECOND ORDER PLOT: $f[OH^-] = 1/[OH^-] - 1/[OH^-]_0$

△ THIRD ORDER PLOT: $f[OH^-] = 1/[OH^-]^2 - 1/[OH^-]_0^2$

$f[OH^-]$

t min

FIG. 9. Plot of data in Fig. 6 according to: ☐, first-order equation (3.5); ○, second-order equation (3.10); △, third-order equation (3.13).

centrations are equal; if they are not, the general forms given in Table 3.4 must be used.

The variation or differential method

The initial rate R of the reaction

$$mA + nB \longrightarrow \text{products}$$

is given by

$$R = k[A]_0{}^m[B]_0{}^n \text{ moles lit}^{-1} \text{ sec}^{-1}$$

where m is the order with respect to A, n is the order with respect to B, and $m + n$ is the overall order.

If various experiments are performed with the same initial concentration $[B]_0$ and different initial concentrations $[A]_0$ their initial rates are given by

$$R = k'[A]_0{}^m \text{ moles lit}^{-1} \text{ sec}^{-1}$$

i.e.

$$R_1/R_2 = ([A_1]_0/[A_2]_0)^m$$

or

$$m = \frac{\log (R_1/R_2)}{\log ([A_1]_0/[A_2]_0)} \quad (3.17)$$

n can be determined in the same way from initial rates measured at different initial concentrations $[B]_0$ with $[A]_0$ constant, and the overall order is obtained by adding the individual orders n and m.

Hinshelwood and Green measured the initial rate R of the gas reaction

$$mNO + nH_2 \longrightarrow \text{products}$$

under the following conditions at 826°C

p_{H_2} (mm)	p_{NO} (mm)	R (mm/sec)	m	n
400	359	150 ⎫		
400	300	103 ⎬ 2·09		
400	152	25 ⎭ 2·09		
289	400	160 ⎫		
205	400	110 ⎬		1·09
147	400	79 ⎭		1·00

m calculated from equation (3.17) with $[A]_0 \propto p_{NO}$ and n obtained in a similar way with $[B]_0 \propto p_{H_2}$ show that the reaction is second-order with respect to NO and first-order with respect to H_2, the overall reaction being third-order in agreement with the data in Fig. 8.

E. Order, Mechanism and Molecularity

The stoichiometric equation

Before any attempt can be made to formulate the mechanism of chemical change, the particular change must be indentified; for example the rate of pyrolysis of an organic vapour can be obtained by measuring the change in pressure and its order established from the pressure–time curves, but *the mechanism by which the reactant undergoes chemical change cannot be considered until its products are identified and their relative amounts measured.* This requires both qualitative and quantitative analysis, often of very small amounts of material in gas phase reactions, which has been facilitated by the development of mass spectrometry and vapour phase chromatography. *The quantitative relationship between reactants and products is expressed by the stoichiometric equation.*

For example, when allowance is made for side reactions it is found that when hydrogen peroxide oxidises hydriodic acid, one mole of iodine is produced for each mole of peroxide consumed; since one mole of iodine requires the oxidation of two moles of hydriodic acid the stoichiometric equation is

$$H_2O_2 + 2HI = 2H_2O + I_2$$

Complex reactions

It is reasonable to assume that molecules can only react when they collide with each other; the rate of reaction should therefore be proportional to the collision frequency which is in turn related to the concentrations of the colliding molecules. If the above reaction takes place in a single step it might therefore be expected that

$$\frac{\text{reaction}}{\text{rate}} \propto \frac{\text{collision frequency of}}{H_2O_2 + 2HI} \propto [H_2O_2][HI]^2$$

i.e. the reaction would be third-order. The experimental rate expression, however, is second-order, which indicates that the rate-controlling process involves a collision between one molecule of each reactant and is probably

$$H_2O_2 + HI \rightarrow H_2O + HIO$$

Since HIO is not a final product and the overall reaction involves an additional molecule of HI, the second stage could be

$$HIO + HI \longrightarrow H_2O + I_2$$

The sum of these elementary processes which constitute the reaction mechanism gives the overall reaction represented by the stoichiometric equation.

The mechanism of decomposition of gaseous N_2O_5

$$2N_2O_5 = 4NO_2 + O_2$$

which is first-order at all but the lowest pressures, is believed to be that outlined in Chapter 1C. In this case the rate-determining step

$$NO_2 + NO_3 \longrightarrow NO_2 + NO + O_2$$

with the velocity

$$k[NO_2][NO_3] \text{ moles lit}^{-1} \text{ sec}^{-1}$$

is so slow that the equilibrium

$$N_2O_5 \rightleftharpoons NO_2 + NO_3$$

is undisturbed and the overall rate becomes

$$-\frac{d[N_2O_5]}{dt} = kK[N_2O_5] \text{ moles lit}^{-1} \text{ sec}^{-1}$$

where

$$K = \frac{[NO_2][NO_3]}{[N_2O_5]} \text{ moles/lit}$$

The mechanism of the gas reaction

$$2NO + 2H_2 = N_2 + 2H_2O$$

which is second-order in nitric oxide and first-order in hydrogen, could be

$$\begin{cases} 2NO + H_2 \longrightarrow N_2 + H_2O_2 \\ H_2O_2 + H_2 \longrightarrow 2H_2O \end{cases}$$

or

$$\begin{cases} 2NO + H_2 \longrightarrow N_2O + H_2O \\ N_2O + H_2 \longrightarrow N_2 + H_2O \end{cases}$$

In either case the second process is much faster than the first which determines the overall rate.

The second-order reaction

$$3KClO = 2KCl + KClO_3$$

is believed to take place in the stages outlined in Chapter 3B.

Simple reactions

The dimerisation of tetrafluoroethylene

$$2CF_2 = CF_2 \rightarrow cyclo - C_4F_8$$

is found to be second-order in the gas phase as would be expected from the stoichiometric equation; consequently this reaction must take place in the single step represented by this equation. This is often the case if the stoichiometric equation contains one or two reactant molecules, whereas if it involves more than two, a complex mechanism should be suspected on the grounds that collisions between more than two molecules are extremely rare.

An apparent exception to this generalisation is afforded by the third-order gas reactions

$$2NO + X_2 = 2NOX$$

where $X = O$, Cl or Br, which presumably take place in a single step as the result of a three-body collision. It has been suggested, however, that nitric oxide reacts in its dimeric form and that the overall reaction is

$$2NO \rightleftharpoons (NO)_2$$
$$(NO)_2 + X_2 \rightarrow 2NOX$$

with

$$-\frac{dp_{NO}}{dt} = kp_{(NO)_2}p_{X_2} \text{ mm/sec}$$

$$= kKp_{NO}^2 p_{X_2} \text{ mm/sec}$$

where

$$K = p_{(NO)_2}/p_{NO}^2 \text{ mm}^{-1}$$

which leads to the observed rate expression. The fact that all known third-order gas reactions involve two molecules of NO supports this suggestion.

Elementary processes

We may conclude from the above discussion that

(a) If the stoichiometric equation predicts the experimentally observed order, the overall reaction probably takes place in a single elementary process represented by this equation.

(b) If the experimental order bears no relation to the stoichiometric equation the overall reaction is complex and involves more than one elementary process.

Elementary processes are the simplest reactions in terms of which an overall chemical change can be represented, and it is with these that theoretical treatments of reaction rates are concerned. It is convenient to classify them according to their *molecularity*, i.e. to *the number of molecules taking part*; the following are examples of unimolecular, bimolecular and termolecular processes, respectively

$$N_2O_5 \longrightarrow NO_2 + NO_3$$
$$KClO + KClO_2 \longrightarrow KCl + KClO_3$$
$$2NO + O_2 \longrightarrow 2NO_2$$

If the overall reaction takes place in one elementary step represented by the stoichiometric equation, e.g.

$$CH_3N_2CH_3 = C_2H_6 + N_2$$
$$2HI = H_2 + I_2$$

it may be referred to as a unimolecular or bimolecular reaction. However, this infers that the mechanism is established and until it is definitely known that the reaction is not complex it is better to refer to the order of an overall reaction.

4

THE VARIATION OF REACTION RATE WITH
TEMPERATURE

A. The Arrhenius Equation

THE rates of most chemical reactions increase enormously as the temperature is raised. An examination of the rate expression in the form

$$\frac{\text{reaction}}{\text{rate}} = \frac{\text{rate}}{\text{constant}} \times \left(\frac{\text{reactant}}{\text{concentration}}\right)^{\text{order}}$$

shows that the rate constant is the temperature-dependent term since

(a) the reactant concentrations are virtually unaffected by temperature;

(b) unless the mechanism changes, in which case the reaction is no longer the same, an increase in temperature will have no effect on the reaction order.

Thus, *although the rate constant is independent of reactant concentration it does vary with temperature* as shown in Fig. 10 for reactions of different types.

Arrhenius (1889) found that the experimentally observed variation of the rate constant k with absolute temperature T^0K can be expressed by the equation

$$\log_e k = B - C/T \tag{4.1}$$

or

$$\frac{d \log_e k}{dT} = \frac{C}{T^2} \tag{4.2}$$

where B and C are constants for the particular reaction. As a test of the Arrhenius equation the logarithms of the rate constants in Fig. 10 are plotted against the reciprocal of the corresponding absolute temperatures in Fig. 11. The excellent straight lines obtained demonstrate the validity of equation (4.1) in each case and are typical of the behaviour of all thermal reactions whether catalysed or uncatalysed,

37

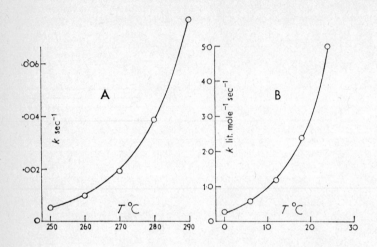

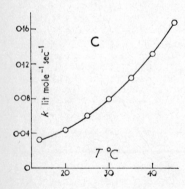

FIG. 10. Variation of rate constant k with temperature.

A. First-order gas reaction
$C_3H_7N_2C_3H_7 = C_6H_{14} + N_2$
(*Ramsperger, 1928*).

B. Second-order reaction in solution
$C_2H_5ONa + CH_3I$
$\quad = C_2H_5OCH_3 + NaI$
(*Hecht and Conrad, 1889*).

C. Acid-catalysed reaction
$CH_3COOH + CH_3OH$
$\quad = CH_3COOCH_3 + H_2O$
(*Williamson and Hinshelwood, 1934*).

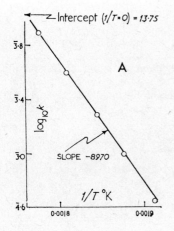

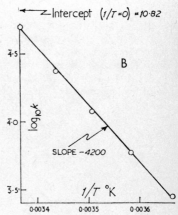

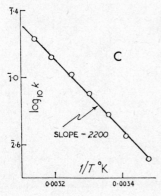

FIG. 11. Plot of data in Fig. 10 according to Arrhenius equation

$$\log_{10} k = \log_{10} A - \frac{E}{2 \cdot 3R} \frac{1}{T}.$$

homogeneous or heterogeneous, or whether they take place in the gas phase or in solution.

B. Activation Energy

The arrangement of atoms in a stable molecule corresponds to a minimum of potential energy. In a chemical reaction the atoms of the reacting molecules corresponding to one potential energy minimum are rearranged to a different configuration, that of the products, which corresponds to another minimum of potential energy. Somewhere in between these minima there must be a potential energy maximum corresponding to an atomic configuration, which the reactants must acquire during the chemical transformation. This is illustrated in Fig. 12.

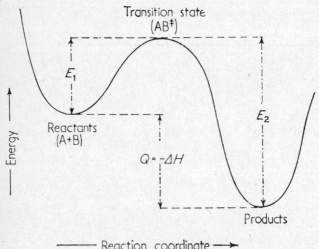

FIG. 12. Variation of potential energy of reactant molecules as they undergo chemical change.

In order to cross this energy barrier the reactants require an energy E_1 and, if the reverse reaction is to take place, the products require an energy E_2. The energies E_1 and E_2, known as the *activation energies* of the forward and reverse reactions, are clearly related (Fig. 12) to the heat Q of the reaction

$$\text{reactants} \underset{k_2}{\overset{k_1}{\rightleftharpoons}} \text{products} + Q \text{ cal/mole}$$

by

$$Q = E_2 - E_1 \text{ cal/mole}$$

or since $Q = -\Delta H$, the change in heat content of the reacting system

$$E_1 - E_2 = \Delta H \text{ cal/mole} \tag{4.3}$$

At equilibrium the rates of forward and reverse reactions are equal, i.e.
$$k_1[\text{reactants}]_e = k_2[\text{products}]_e \text{ moles lit}^{-1} \text{ sec}^{-1}$$

and the equilibrium concentrations $[\text{reactants}]_e$ and $[\text{products}]_e$ are related by

$$\frac{[\text{products}]_e}{[\text{reactants}]_e} = \frac{k_1}{k_2} = K \tag{4.4}$$

where the equilibrium constant K varies with temperature according to the thermodynamic expression

$$\frac{\mathrm{d} \log_e K}{\mathrm{d}T} = \frac{\Delta H}{RT} \tag{4.5}$$

Substituting for K from (4.4) and for ΔH from (4.3), equation (4.5) becomes

$$\frac{\mathrm{d} \log_e (k_1/k_2)}{\mathrm{d}T} = \frac{\mathrm{d} \log_e k_1}{\mathrm{d}T} - \frac{\mathrm{d} \log_e k_2}{\mathrm{d}T} = \frac{E_1 - E_2}{RT^2}$$

i.e.
$$\frac{\mathrm{d} \log_e k_1}{\mathrm{d}T} = \frac{E_1}{RT^2} + J$$

$$\frac{\mathrm{d} \log_e k_2}{\mathrm{d}T} = \frac{E_2}{RT^2} + J$$

which are identical with the Arrhenius empirical expression (4.2) if the constant $J = 0$ and $C = E/R$.

The Arrhenius equation is usually written:

$$k = A e^{-E/RT} \tag{4.6}$$

where $A = $ antilog B. This form conveniently expresses the temperature dependence of k in terms of

(a) a pre-exponential or frequency factor A which has the same dimensions as the rate constant itself;

(b) an activation energy E which is usually given in calories per mole.

Both **A** *and* **E** *are independent of temperature to a first approximation and both are of great theoretical significance in that they are determined by the properties of the reacting molecules themselves.*

Clearly once A and E are known for the reaction

$$C + D \longrightarrow \text{products}$$

its rate may be calculated for any conditions of reactant concentration and temperature from the rate expression

$$-\frac{d[C]}{dt} = Ae^{-E/RT}[C][D] \text{ moles lit}^{-1} \text{ sec}^{-1}$$

C. The Experimental Determination of A and E

The logarithmic form of equation (4.6) is

$$\log_e k = \log_e A - E/RT$$

or since it is more convenient to use $\log_{10}$

$$\log_{10} k = \log_{10} A - E/2\cdot303\ RT \qquad (4.7)$$

If k is calculated from the concentration–time curves obtained at different temperatures then a plot of $\log_{10} k$ against $1/T$ should be linear with (4.7)

(a) a slope of $-E/2\cdot303\ R$;

(b) an intercept on the $\log_{10} k$ axis ($1/T = 0$) of $\log_{10} A$. Such plots are illustrated in Fig. 11; for the first-order decomposition of azoisopropane vapour Fig. 11A gives

slope $= -8950$

i.e. $\qquad E = -(-8950) \times 2\cdot303 \times 1\cdot98 = 41000 \text{ cal/mole}$

intercept $= 13\cdot75$

i.e. $\qquad A = 5\cdot6 \times 10^{13} \text{ sec}^{-1}$

The complete rate expression for this reaction is therefore

$$-\frac{d[C_3H_7N_2C_3H_7]}{dt}$$

$$= 5\cdot6 \times 10^{13}e^{-41000/RT}[C_3H_7N_2C_3H_7] \text{ moles lit}^{-1} \text{ sec}^{-1}$$

Table 4.1 lists experimental values of A and E for a number of other reactions.

TABLE 4.1. Some Arrhenius Parameters

Unimolecular reactions	E (kcal/mole)†	A (sec^{-1})
$F_2O_2 = F_2 + O_2$	17·3	6×10^{12}
$C_2H_5Cl = C_2H_4 + HCl$	60·8	4×10^{14}
$C_2H_5Br = C_2H_4 + HBr$	52·3	7×10^{12}
$CH_3COOC_2H_5 = CH_3COOH + C_2H_4$	47·8	3×10^{12}
	23·0	1×10^{14}
	43·0	6×10^{12}
	33·7	1×10^{13}

Bimolecular reactions	E (kcal/mole)	A (lit mole^{-1} sec^{-1})
$H_2 + I_2 = 2HI$	39·5	$1·6 \times 10^{11}$
$2HI = H_2 + I_2$	44·5	$9·2 \times 10^{10}$
$2NO_2 = 2NO + O_2$	26·9	$9·4 \times 10^9$
$2NOBr = 2NO + Br_2$	13·9	$4·2 \times 10^{10}$
$C_2H_4 + H_2 = C_2H_6$	43·1	$1·2 \times 10^6$
$CH_3COOC_2H_5 + OH^- = CH_3COO^- + C_2H_5OH$	14·7	$7·9 \times 10^5$
$CH_3I + (C_2H_5)_3N = CH_3N(C_2H_5)_3I^-$	9·7	$2·1 \times 10^4$

† 1 kilocalorie (kcal) = 1000 calories

5

THE THEORETICAL TREATMENT OF REACTION RATES

THE analysis of experimental data discussed in the previous chapters may be summarised as follows:—

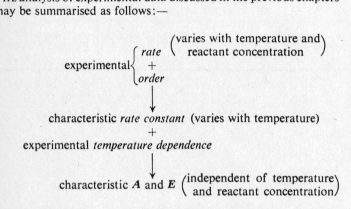

$$\text{experimental}\begin{cases} rate \\ + \\ order \end{cases} \left(\begin{array}{c} \text{varies with temperature and} \\ \text{reactant concentration} \end{array}\right)$$

$$\downarrow$$

characteristic *rate constant* (varies with temperature)
+
experimental *temperature dependence*

$$\downarrow$$

characteristic A and E $\left(\begin{array}{c} \text{independent of temperature} \\ \text{and reactant concentration} \end{array}\right)$

The Arrhenius parameters A and E are determined by the properties of the reacting molecules themselves, consequently any theoretical approach to an understanding of reaction rates must be concerned with the calculation of A and E from these properties; once A and E are known it is a simple matter to predict reaction rates for any conditions of temperature and reactant concentration. The theoretical treatment is simplified by considering elementary reactions since these are well-defined collisional processes and their rates are in any case related to the rate of the overall reaction.

At the present time the approximations involved in calculating activation energies are such as to remove any confidence in the results; however, a good deal of progress has been made in our understanding of the pre-exponential A factor in terms of the theories outlined below.

44

A. Theory of Unimolecular Reactions

Mechanism

Although a unimolecular process involves only one reactant molecule this must clearly acquire the necessary energy of activation in a bimolecular collision and the overall reaction should be second-order. The fact that unimolecular reactions are first-order except in the gas phase at low pressures has been explained by Lindemann (1922), who pointed out that if an activated molecule does not undergo an immediate chemical change it may be collisionally deactivated instead. A unimolecular gas reaction can therefore be represented by the following processes:—

$$
\begin{array}{lll}
& & Rate \\
(1) & M + M \longrightarrow M^* + M & k_1[M]^2 \\
(2) & M^* + M \longrightarrow M + M & k_2[M^*][M] \\
(3) & \qquad M^* \longrightarrow N & k_3[M^*]
\end{array}
$$

where the asterisk denotes an activated reactant molecule M and only process (3) involves a chemical change.

If we assume that the rate of production of activated molecules is equal to the rate of removal of activated molecules, i.e.

$$k_1[M]^2 = k_2[M^*][M] + k_3[M^*]$$

then
$$[M^*] = \frac{k_1[M]^2}{k_2[M] + k_3}$$

and the rate of reaction is

$$\frac{d[N]}{dt} = k_3[M^*] = \frac{k_1 k_3 [M]^2}{k_2[M] + k_3} \tag{5.1}$$

At sufficiently high pressures of reacting gas, the bimolecular process (2) will be much faster than the unimolecular change (3), i.e.

$$k_2[M] \gg k_3$$

in which case (5.1) reduces to the first-order expression

$$\frac{d[N]}{dt} = \frac{k_1 k_3 [M]}{k_2} \tag{5.2}$$

which represents the behaviour observed.

If on the other hand the pressure of reacting gas is so low that the probability of collisional deactivation is negligible compared with the probability of reaction, i.e.

$$k_2[M] \ll k_3$$

the collisional activation process (1) becomes rate-controlling and (5.1) reduces to the second-order expression

$$\frac{d[N]}{dt} = k_1[M]^2 \tag{5.3}$$

At intermediate pressures the reaction rate is given by (5.1) and the order is between first and second.

The change from first- to second-order behaviour at low pressures is characteristic of unimolecular gas reactions and strikingly confirms the Lindemann mechanism. If the activation energy must be associated with some specific vibration of a polyatomic molecule, it is readily understood that activation of the molecule as a whole is not a sufficient condition for reaction and that the internal redistribution of this energy is responsible for the time-lag between activation and reaction. As a rule this time-lag ($= 1/k_3$) increases with molecular complexity so that the actual pressure at which the change in order is observed is higher the smaller the reacting molecule; diatomic molecules dissociate so rapidly following collisional activation that the probability of deactivation is negligible and the dissociation is second-order up to the highest gas pressures.

In solution where the reacting molecule experiences very frequent collisions with solvent molecules, the unimolecular change is always rate-controlling and first-order kinetics are observed regardless of the reactant concentration.

The A factor

A chemical change involves the rupture and formation of chemical bonds; consequently it is reasonable to suppose that for a unimolecular process

(a) *the activation energy E must be acquired as vibrational energy of the atoms to be separated;*

(b) *once this energy requirement is met, the rate of bond rupture is given by the corresponding vibrational frequency v sec^{-1}.*

If n is the number of reactant molecules per cubic centimetre, the concentration of molecules with energy not less than E associated with a particular vibration is

$$n_E = n e^{-E/RT} \text{ molecules/cm}^3$$

and the rate of reaction is

$$-\frac{dn}{dt} = v n_E = v n e^{-E/RT} \text{ molecules cm}^{-3} \text{ sec}^{-1}$$

$$= kn \text{ molecules cm}^{-3} \text{ sec}^{-1}$$

whence $k = v \mathrm{e}^{-E/RT} \sec^{-1}$ (5.4)

A comparison of (5.4) with the Arrhenius equation (4.6) shows that the A factor for unimolecular processes should be equal to a particular vibrational frequency of the reacting molecule which is usually of the order of 10^{12}–10^{14} sec^{-1}; an examination of the A factors for unimolecular reactions, of which those given in Table 4.1 are typical, shows that this is the case.

B. Collision Theory of Bimolecular Gas Reactions

The application of the simple kinetic theory of gases to the rate of the bimolecular process

$$A + B \longrightarrow \text{products}$$

is based on the reasonable assumption that for the reaction to take place

(a) *molecules* A *and* B *must collide.*

(b) *the energy of the colliding molecules must be at least equal to the activation energy* E.

The rate is then equal to the number of effective collisions per second in unit volume of the reacting gas.

The probability W_A that molecule A has energy at least equal to ε at the absolute temperature T is

$$W_A = \mathrm{e}^{-\varepsilon/RT}$$

whilst the probability W_B that molecule B has energy not less than $(E - \varepsilon)$ at the same temperature is

$$W_B = \mathrm{e}^{-(E-\varepsilon)/RT}$$

The probability that molecules A and B collide with a combined energy at least equal to E is therefore

$$W_A W_B = \mathrm{e}^{-\varepsilon/RT}\mathrm{e}^{-(E-\varepsilon)/RT} = \mathrm{e}^{-E/RT}$$

which is equal to the fraction of effective collisions when large numbers of molecules are involved; i.e.

$$\frac{\text{number of effective collisions/cm}^3 \text{ per sec}}{\text{total number of collisions/cm}^3 \text{ per sec}} = \mathrm{e}^{-E/RT} \quad (5.5)$$

If the concentrations of A and B are n_A and n_B molecules/cm^3, respectively, then

$$\text{total number of collisions/cm}^3 \text{ per sec} = Z n_A n_B \quad (5.6)$$

where Z, the collision number, is equal to the collision frequency when the concentrations of A and B are one molecule/cm^3. Since

$$\text{reaction rate} = \text{number of effective collisions/cm}^3 \text{ per sec}$$

equations (5.5) and (5.6) lead to

$$\text{reaction rate} = Zn_A n_B e^{-E/RT} \text{ molecules cm}^{-3} \text{ sec}^{-1}$$
$$= kn_A n_B \text{ molecules cm}^{-3} \text{ sec}^{-1}$$

whence $\qquad k = Z e^{-E/RT} \text{ cm}^3 \text{ molecule}^{-1} \text{ sec}^{-1}$ (5.7)

According to the simple collision theory *the A factor of the Arrhenius expression* (4.6) *for bimolecular reactions is equal to the collision number Z*, which can be calculated from the kinetic theory expression

$$Z = (\sigma_A + \sigma_B)^2 [8\pi RT(M_A + M_B)/M_A M_B]^{\frac{1}{2}}$$ (5.8)

if the collision radii σ_A and σ_B and molecular weights M_A and M_B of reactants A and B are known. (It should be noted that Z and hence A increases as $\sqrt{T}$ since the relative velocity of the molecules is proportional to $\sqrt{T}$; however, this temperature dependence of A may be considered negligible in comparison with the temperature sensitivity of the term $e^{-E/RT}$ in the rate expression.)

As a test of the simple collision theory values of Z calculated from equation (5.8) are compared with the experimental A factors for a number of bimolecular gas reactions in Table 5.1. The good agreement between these quantities for reactions involving simple molecules suggests that conditions (a) and (b) are sufficient in these cases. For reactions between larger molecules where A is less than Z by at least a factor of 10, it appears that the additional condition (c) must also be satisfied,

(c) *The molecules must be suitably oriented on collision.*

This condition can be taken into account by the introduction of a *steric or probability factor P* into (5.7) such that

$$k = PZ e^{-E/RT} \text{ cm}^3 \text{ molecule}^{-1} \text{ sec}^{-1}$$ (5.9)

where $P = A/Z \leqslant 1$. The very large range of P values obtained is illustrated by the data in Table 5.1.

Although this simple model provides an understanding of the factors involved in the A factor for bimolecular processes, it gives no information concerning the magnitude of the probability factor P which can only be calculated from the experimental value of A.

C. The Transition State Theory

This approach treats the reaction in terms of an equilibrium between reactants A and B and the *transition state $AB^{\ddagger}$* at the top of the potential energy barrier shown in Fig. 12, i.e.

$$A + B \rightleftharpoons AB^{\ddagger} \longrightarrow \text{products}$$

TABLE 5.1. Steric Factors for Bimolecular Gas Reactions

Reaction	A (lit mole^{-1}sec^{-1})	Z (lit mole^{-1}sec^{-1})	P ($= A/Z$)
$2HI = H_2 + I_2$	4.6×10^{10}	4.6×10^{10}	1.0
$2NOCl = 2NO + Cl_2$	7.3×10^{10}	7.2×10^{10}	1.0
$H_2 + I_2 = 2HI$	1.5×10^{11}	5×10^{11}	0.3
$2NO_2 = 2NO + O_2$	9.4×10^{9}	1.6×10^{11}	6×10^{-2}
$2 \; CF_2{=}CF_2 = \begin{array}{c} CF_2{-}CF_2 \\ | \qquad | \\ CF_2{-}CF_2 \end{array}$	1.6×10^{8}	2×10^{11}	8×10^{-4}
$\begin{array}{c} CH_2 \\ \| \qquad CH{=}CH \\ CH \\ \| \\ CH \\ \| \qquad CH{=}CH \\ CH_2 \end{array} + \begin{array}{c} CH_2 \\ \| \\ CH_2 \end{array} = \begin{array}{c} CH_2 \qquad CH{=}CH \\ | \qquad \qquad | \\ CH \qquad CH_2 \\ | \qquad \qquad | \\ CH_2 \qquad CH_2 \\ \quad CH_2{-}CH_2 \end{array}$	3×10^{7}	7.5×10^{11}	4×10^{-5}
$\begin{array}{c} CH \\ \| \qquad CH{=}CH \\ CH \\ \| \\ CH \\ \| \qquad CH{=}CH \\ CH \end{array} = (\text{bicyclic product structure})$	1.3×10^{6}	3.4×10^{11}	4×10^{-6}

The transition state or activated complex $AB^{\ddagger}$ is regarded as a normal molecule in every respect except that one of its vibrations is equivalent to a translational degree of freedom along the reaction coordinate which leads to the formation of products. If this degree of freedom is treated as a classical vibration, its frequency ν at temperature T is given by

$$h\nu = kT \text{ ergs/molecule}$$

where h is Planck's constant and $k = R/N$ ergs molecule^{-1} °C^{-1} is the gas constant per molecule.

The reaction rate according to this model is equal to the concentration of activated complexes multiplied by the frequency with which they cross the barrier to form products, i.e.

$$\text{reaction rate} = \nu[AB^{\ddagger}] = \frac{kT}{h}[AB^{\ddagger}]$$

$$= k[A][B]$$

whence

$$k = \frac{kT}{h}\frac{[AB^{\ddagger}]}{[A][B]} = \frac{kT}{h}K^{\ddagger} \qquad (5.10)$$

$K^{\ddagger}$ is the equilibrium constant for the process

$$A + B \rightleftharpoons AB^{\ddagger}$$

and like any other equilibrium constant is related to the free energy of formation $\Delta G^{\ddagger}$ of the activated complex by the thermodynamic expression

$$RT \log_e K^{\ddagger} = -\Delta G^{\ddagger}$$

or

$$K^{\ddagger} = e^{-\Delta G^{\ddagger}/RT}$$

whence

$$k = \frac{kT}{h}e^{-\Delta G^{\ddagger}/RT} \qquad (5.11)$$

and the rate of reaction is controlled by the free energy of activation $\Delta G^{\ddagger}$.

Since $\Delta G^{\ddagger}$ is related to the heat of activation $\Delta H^{\ddagger}$ and entropy of activation $\Delta S^{\ddagger}$ by the thermodynamic relationship

$$\Delta G^{\ddagger} = \Delta H^{\ddagger} - T\Delta S^{\ddagger}$$

(5.11) may be written

$$k = \frac{kT}{h}e^{\Delta S^{\ddagger}/R}e^{-\Delta H^{\ddagger}/RT} \qquad (5.12)$$

which may be compared with the collision theory expression

$$k = PZe^{-E/RT}$$

Thus assuming $E = \Delta H^{\ddagger}$

$$P = \frac{kT}{Zh} e^{\Delta S^{\ddagger}/R}$$

i.e. the steric or probability factor P is related to the entropy of activation $\Delta S^{\ddagger}$.

The advantages of the transition state theory are

(a) it provides a quantitative treatment of the steric factor in bimolecular reactions;

(b) it is not restricted to bimolecular processes, equation (5.12) being equally applicable to the rate constants of unimolecular and termolecular processes;

(c) it can be applied to reactions in solution where the collision number Z is not strictly that obtained from the simple gas kinetic expression (5.8).

Although in principle the entropy of activation can be calculated from vibrational frequencies, bond lengths and other properties of the reactants and of the complex, the values obtained are very approximate, since these properties of the transition state are not known with certainty. In the case of unimolecular reactions of the type

$$B \rightleftharpoons B^{\ddagger} \longrightarrow C + D$$

the reactant B and activated complex $B^{\ddagger}$ should be very similar in structure and $\Delta S^{\ddagger} \approx 0$; consequently

$$A \approx kT/h \approx 10^{13} \text{ sec}^{-1}$$

which is found to be the case in a large number of unimolecular processes. On the other hand, the bimolecular addition reaction

$$B + C \rightleftharpoons BC^{\ddagger} \longrightarrow D$$

proceeds via an activated complex $BC^{\ddagger}$ which must have approximately the same entropy as the final product D; in this case the entropy of activation

$$\Delta S^{\ddagger} = S_{BC^{\ddagger}} - (S_B + S_C)$$

should be very nearly equal to the overall entropy of reaction

$$\Delta S = S_D - (S_B + S_C)$$

The qualitative agreement between experimental A factors for reactions of this type and A factors calculated from

$$A = \frac{kT}{h} e^{\Delta S/R}$$

on the assumption that $\Delta S^{\ddagger} = \Delta S$ shows that this is the case.

Finally it is emphasised that the entropy and heat of *activation* $\Delta S^{\ddagger}$ and $\Delta H^{\ddagger}$, which determine the reaction rate, *must not be confused* with the entropy and enthalpy changes of the *reaction* ΔS and ΔH which dictate the extent of reaction.

6

HOMOGENEOUS CATALYSIS

A. Characteristics of Catalysed Reactions

A CATALYST increases the rate of chemical reaction. Since the addition of a catalyst does not change the stoichiometry of the reaction

$$A + B \longrightarrow C + D$$

otherwise the reaction would not be the same one, the catalysed reaction may be represented by

$$A + B + \text{catalyst} \longrightarrow C + D + \text{catalyst}$$

from which it follows that

(a) *the catalyst remains chemically unchanged at the end of the reaction.*

As a consequence of its unchanged concentration, the free energy of the catalyst is also unchanged and the change in free energy of the reaction ΔG is unaffected by its presence; therefore according to the thermodynamic expression

$$\Delta G = -RT \log_e K$$

(b) *a catalyst does not alter the position of chemical equilibrium.* Since at equilibrium the ratio of forward and reverse reactions are equal it follows that

(c) *a catalyst increases the rates of both forward and reverse reactions to the same extent*, in view of which it is better defined as *a substance which accelerates the approach to chemical equilibrium.* Because the catalyst is not consumed during the reaction it is clear that

(d) *a small amount of catalyst will promote a large amount of chemical change.*

The transition state theory leads to the expression

$$k = \frac{kT}{h} e^{-\Delta G^{\ddagger}/RT} = \frac{kT}{h} e^{\Delta S^{\ddagger}/R} e^{-\Delta H^{\ddagger}/RT}$$

for the rate constant k; since k, h and R are universal constants and the addition of a catalyst does not change the temperature T, the increased rate of a catalysed reaction must be due to a reduction in the free energy of activation $\Delta G^{\ddagger}$. This is largely brought about by a reduction in the energy of activation $E(\approx \Delta H^{\ddagger})$ as shown by the data for the following gas reactions which are catalysed by iodine vapour:

Pyrolysis of	E (uncatalysed)	E (catalysed)
$C_2H_5OC_2H_5$	53 500 cal/mole	34 300 cal/mole
CH_3CHO	45 500 cal/mole	32 500 cal/mole

If the activation energy of an uncatalysed reaction is reduced from E_1 to $(E_1 - \Delta E)$ in the presence of a catalyst, the activation energy E_2 of the reverse process is reduced to E_2' where

$$E_2' = (E_1 - \Delta E) - \Delta H$$
$$= E_2 - \Delta E$$

since ΔH remains unchanged and $(E_1 - \Delta H) = E_2$. The catalyst therefore reduces the activation energies of both forward and

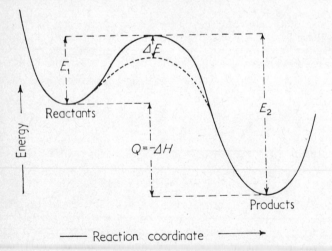

FIG. 13. Reaction paths for uncatalysed reaction (solid line) and catalysed reaction (dashed line) in which activation energy is lowered by ΔE. (N.B. The heat of reaction Q is unchanged.)

reverse reactions by the same amount, i.e. it provides an alternative reaction path with a lower energy barrier as shown in Fig. 13.

It is convenient to classify catalysis as *homogeneous* if the catalyst and reactants form a single phase as in the iodine-catalysed gas reactions above, or as *heterogeneous* if the catalyst constitutes an additional phase as does the platinum gauze in the gaseous oxidation of ammonia. Heterogeneous catalysis is discussed in the following chapter.

B. The Kinetics of Homogeneous Catalysis

The overall rate of the homogeneous catalysed reaction

$$A + B + \text{catalyst} \longrightarrow C + D + \text{catalyst}$$

is strictly the sum of the rates of catalysed and uncatalysed processes, i.e.

$$-\frac{\mathrm{d}[A]}{\mathrm{d}t} = k[A][B] + k_c[A][B][\text{catalyst}]$$

but since the uncatalysed reaction is usually very much slower than the catalysed process this rate expression may be reduced to

$$-\frac{\mathrm{d}[A]}{\mathrm{d}t} = k_c[A][B][\text{catalyst}] \tag{6.1}$$

The rate constant k_c of the catalysed reaction, often referred to as the

TABLE 6.1. Effect of Concentration of Acetic Acid Catalyst [HA] on Rate of Reaction $CH_3CH(OH)_2 = CH_3CHO + H_2O$

(Bell and Higginson, 1949)

[HA] (moles/lit)	k (min^{-1})	$k_{HA} = k/[HA]$ (lit mole^{-1} min^{-1})
0·00101	0·0201	20·0
0·00148	0·0308	20·8
0·00195	0·0390	20·0
0·00220	0·0435	19·8
0·00289	0·0585	20·2
0·00295	0·0565	19·2
0·00383	0·0783	20·4
0·00450	0·0855	19·0
0·00550	0·1100	20·0

catalytic constant or catalytic coefficient, is a measure of the efficiency of the catalyst for the particular reaction. Since the concentration of catalyst remains unchanged, equation (6.1) may be written

$$-\frac{\mathrm{d}[A]}{\mathrm{d}t} = k_c'[A][B] \tag{6.2}$$

where $k_c' = k_c$[catalyst], so that the catalysed reaction behaves kinetically as the uncatalysed one. However, the measured rate constant k_c' varies with the catalyst concentration which must be known before the catalytic constant k_c can be evaluated. This is illustrated by the data in Table 6.1.

C. Catalysis by Acids and Bases

One of the most important examples of homogeneous catalysis in solution is that exhibited by acids and bases. The hydrolyses of esters, amides and of diazoacetic ester

$$CH_3COOC_2H_5 + H_2O \xrightarrow{H^+} CH_3COOH + C_2H_5OH$$

$$CH_3CONH_2 + H_2O \xrightarrow{H^+} CH_3COOH + NH_3$$

$$N_2CHCOOC_2H_5 + H_2O \xrightarrow{H^+} HOCH_2COOC_2H_5 + N_2$$

are catalysed by acids, whilst an example of base-catalysed processes is the dissociation of diacetone alcohol

$$(CH_3)_2C(OH)CH_2COCH_3 \xrightarrow{OH^-} 2CH_3COCH_3$$

It is found that for a given acid-catalysed reaction, the catalytic constant k_{HA} determined from

$$\text{reaction rate} = k_{HA}[\text{reactant}][HA]$$

increases with the strength of the acid HA. If, however, the acid concentration is replaced by the concentration of hydrogen ions such that

$$\text{reaction rate} = k_{H^+}[\text{reactant}][H^+]$$

the catalytic constant k_{H^+} is the same for all acids for the same reaction. This is illustrated by the data in Table 6.2. Since the undissociated acid also exhibits a catalytic effect, however, the rate expression is strictly:

$$\text{reaction rate} = k_{H^+}[\text{reactant}][H^+] + k_{HA}[\text{reactant}][HA]$$
$$= k[\text{reactant}]$$

TABLE 6.2. Effect of Acid Concentration [HA] on the Rate of Hydrolysis of Diazoacetic Ester

$$N_2CHCOOC_2H_5 + H_2O = HOCH_2COOC_2H_5 + N_2$$

(*Duboux and Pièce, 1940*)

[HA] (moles/lit)	k (min^{-1})	$k_{HA} = k/[HA]$ (lit mole^{-1} min^{-1})	$k_H = k/[H^+]$ (lit mole^{-1} min^{-1})
		HA = acetic acid	
0·00098	0·00457	4·66	36·9
0·00375	0·00929	2·48	37·0
0·01820	0·02180	1·20	38·8
0·04885	0·03614	0·74	38·9
		HA = benzoic acid	
0·00099	0·000219	0·221	37·0
0·00200	0·000324	0·162	37·3
0·00325	0·000426	0·131	37·3
0·00992	0·000790	0·080	39·0
		HA = succinic acid	
0·00050	0·00535	10·70	34·7
0·00099	0·00831	8·39	36·3
0·00500	0·02080	4·16	38·0
0·01000	0·03030	3·03	38·5

where the measured rate constant k is given by

$$k = k_{H^+}[H^+] + k_{HA}[HA]$$

This behaviour in which each acid constituent contributes to the overall catalytic effect is known as *generalised acid catalysis*. In the same way the measured rate constant of a reaction undergoing *generalised base catalysis* is equal to the sum of the concentrations of basic species present each multiplied by its appropriate catalytic constant.

Some reactions such as the mutarotation of glucose and the enolisation of acetone

$$CH_3COCH_3 \rightarrow CH_3C{=}CH_2$$
$$\underset{\underset{OH}{|}}{}$$

are catalysed by both acids and bases. These are subject to *generalised acid–base catalysis*, the measured rate constant in the presence of acid HA being given by

$$k = k_{H^+}[H^+] + k_{HA}[HA] + k_{A^-}[A^-]$$

and in the presence of base B^- by

$$k = k_{B^-}[B^-] + k_{BH}[BH] + k_{H^+}[H^+]$$

where A^- is the conjugate base of acid HA and BH is the conjugate acid of base B^-.

The mechanism of acid-catalysed ester hydrolysis is believed to be

whilst the base-catalysed dimerisation of acetone is understood to take place in the following stages

D. Intermediate Compound Theory of Homogeneous Catalysis

A characteristic of homogeneous catalysis is the dependence of reaction rate on catalyst concentration (6.1) This suggests that the catalyst is playing a chemical role. Since, however, the overall concentration of catalyst remains unchanged, any compound formed between reactants and catalyst must react in a subsequent process to regenerate the catalytic species, and may therefore be regarded as a reaction intermediate.

Examples of intermediate compound formation are provided by the charged complexes taking part in the acid- and base-catalysed reactions outlined above. In the gas phase the nitric oxide catalysed oxidation of sulphur dioxide and of carbon monoxide are thought to be due to the complex reactions

$$NO + \tfrac{1}{2}O_2 \rightarrow NO_2$$
$$NO_2 + SO_2 \rightarrow NO + SO_3$$

and
$$NO + \tfrac{1}{2}O_2 \rightarrow NO_2$$
$$NO_2 + CO \rightarrow NO + CO_2$$

where nitrogen dioxide is the intermediate compound in each case. The addition of iodine vapour to acetaldehyde at moderate temperatures is believed to change the reaction mechanism from the simple unimolecular process

$$CH_3CHO \rightarrow CH_4 + CO$$

to the composite reaction

$$CH_3CHO + I_2 \rightarrow CH_3I + HI + CO$$
$$CH_3I + HI \rightarrow CH_4 + I_2$$

whilst the Friedel–Crafts reaction probably involves the formation of an intermediate complex between the aluminium chloride catalyst and the acid chloride

$$CH_3COCl + AlCl_3 \rightarrow (CH_3CO^+)(AlCl_4{}^-)$$
$$CH_3CO^+ + C_6H_6 \rightarrow CH_3COC_6H_5 + H^+$$
$$H^+ + AlCl_4{}^- \rightarrow AlCl_3 + HCl$$

E. Catalysis of Ionic Reactions in Solution—the Primary Salt Effect

Since ions exert considerable electrostatic forces on each other at distances much greater than their collision diameters they cannot be treated as ideal solute species. The active concentration of an ion is strictly given by its *activity a* which is the product of its actual concentration c and an activity coefficient f which is less than unity except in infinitely dilute solutions, i.e.

$$a = cf \leqslant c$$

According to the transition state theory the rate of the reaction

$$A + B \rightleftharpoons AB^{\ddagger} \longrightarrow \text{products}$$

is
$$-\frac{d[A]}{dt} = \frac{kT}{h}[AB^{\ddagger}]$$

$$= k[A][B]$$

k, the experimental rate constant, is measured in terms of concentrations $[A]$ and $[B]$ to which it is related by

$$k = \frac{kT}{h} \frac{[AB^{\ddagger}]}{[A][B]}$$

If A and B are ions, the equilibrium constant $K^{\ddagger}$ for the formation of activated complex $AB^{\ddagger}$ is strictly given by

$$K^{\ddagger} = \frac{a_{AB^{\ddagger}}}{a_A a_B} = \frac{[AB^{\ddagger}]}{[A][B]} \frac{f_{AB^{\ddagger}}}{f_A f_B}$$

whence
$$k = \frac{kT}{h} K^{\ddagger} \frac{f_A f_B}{f_{AB^{\ddagger}}} = k_0 \frac{f_A f_B}{f_{AB^{\ddagger}}} \tag{6.3}$$

if k_0 is the rate constant at infinite dilution when $f_A = f_B = f_{AB^{\ddagger}} = 1$. Thus any added substance which changes the activity coefficients of the reactants and transition state will change the measured rate of reaction.

In dilute solutions the activity coefficient f of an ion of charge z is related to the *ionic strength* μ of the solution by the Debye–Huckel expression

$$-\log f = 0\cdot 5 z^2 \sqrt{\mu}$$

so that
$$\log \frac{f_A f_B}{f_{AB}} = -0\cdot 5 \sqrt{\mu}(z_A^2 - z_B^2 - z_{AB^{\ddagger}}^2) \tag{6.4}$$

Since the charge on the activated complex $z_{AB^{\ddagger}}$ is equal to the algebraic sum of the ionic charges z_A and z_B, i.e.

$$z_{AB^{\ddagger}} = z_A + z_B$$

equation (6.4) becomes

$$\log \frac{f_A f_B}{f_{AB}} = z_A z_B \sqrt{\mu}$$

which may be substituted in the logarithmic form of equation (6.3)

$$\log k = \log k_0 + \log \frac{f_A f_B}{f_{AB^{\ddagger}}}$$

to give
$$\log k = \log k_0 + z_A z_B \sqrt{\mu} \tag{6.5}$$

The validity of expression (6.5), known as the Brönsted–Bjerrum equation after its originators, is illustrated in Fig. 14 where $\log (k/k_0)$ is plotted against ionic strength for a number of reactions between ions; in each case the slope is equal to the product of the charges of the reacting ions $z_A z_B$. The slope of -1 obtained for the reaction

$$NH_4CNO \longrightarrow CO(NH_2)_2$$

indicates that the reactants in this case are the ions NH_4^+ ($z = +1$) and CNO^- ($z = -1$), and that the reaction is bimolecular

$$NH_4^+ + CNO^- \longrightarrow CO(NH_2)_2$$

The ionic strength μ of a solution is half the sum of the individual ionic concentrations each multiplied by the square of the ionic charge, i.e.

$$\mu = \tfrac{1}{2}c_1z_1^2 + \tfrac{1}{2}c_2z_2^2 + \tfrac{1}{2}c_3z_3^2 + \ldots$$

where c_1, c_2, c_3 ... and z_1, z_2, z_3 ... are the concentrations and charges of ionic species 1, 2, 3 ... The ionic strength of a solution of $NaNO_3$ at concentration c moles/lit is

$$\mu_{NaNO_3} = \tfrac{1}{2}(c)(+1)^2 + \tfrac{1}{2}(c)(-1)^2 = \tfrac{1}{2}c + \tfrac{1}{2}c = c$$

but if either ion has a multiple charge the ionic strength is greater than the salt concentration, e.g.

$$\mu_{CaCl_2} = \tfrac{1}{2}(c)(+2)^2 + \tfrac{1}{2}(2c)(-1)^2 = 2c + c = 3c$$

According to the Brönsted–Bjerrum equation (6.5) the addition of any salt will have a marked effect on the rate of a reaction between ions in solution even though it exerts no chemical effect; if the reacting ions have the *same* charge $z_A z_B$ is positive and the addition of a salt will *increase* the slow reaction rate, but if the ions are of *opposite* charge $z_A z_B$ is negative and the much faster reaction rate is *reduced* by the increase in ionic strength due to the added salt.

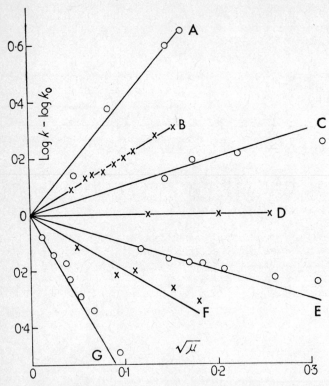

FIG. 14. Variation of rate constant k with ionic strength μ for ionic reactions illustrating validity of Brönsted–Bjerrum equation (6.5).

Reaction	Slope $= Z_A Z_B$
A. $2[CoBr(NH_3)_5]^{++} + Hg^{++} + 2H_2O$ $= 2[CoH_2O(NH_3)_5]^{+++} + HgBr_2$ (*Brönsted and Livingston, 1927*)	$+4$
B. $S_2O_8^- + 2I^- = 2SO_4^- + I_2$ (*King and Jacobs, 1931*)	$+2$
C. $CO(OC_2H_5)N:NO_2^- + OH^- = C_2H_5OH + N_2O + CO_3^-$ (*Brönsted and Delbanco, 1925*)	$+1$
D. $[Cr(urea)_6]^{+++} + 6H_2O = [Cr(H_2O)_6]^{+++} + 6$ urea (*Kilpatrick, 1928*)	0
E. $H_2O_2 + 2H^+ + 2Br^- = 2H_2O + Br_2$ (*Livingston, 1926*)	-1
F. $[CoBr(NH_3)_5]^{++} + OH^- = [CoOH(NH_3)_5]^{++} + Br^-$ (*Brönsted and Livingston, 1927*)	-2
G. $Fe^{++} + Co(C_2O_4)_3^{\equiv} = Fe^{+++} + 3C_2O_4^- + Co^{++}$ (*Barnett and Baxendale, 1956*)	-6

7

HETEROGENEOUS CATALYSIS

A. Characteristics of Heterogeneous Catalysis

A HETEROGENEOUS process takes place at a phase boundary, e.g. at a solid surface. If the same reaction can take place homogeneously, the heterogeneous process is in this sense an example of heterogeneous catalysis. Such processes include

(a) the hydrogenation of unsaturated organic liquids at the surface of a finely divided nickel catalyst;

(b) the catalytic cracking (breaking down) of high molecular weight crude oil to produce volatile hydrocarbons of low molecular weight used as fuel in internal combustion engines;

(c) fermentation and physiological processes catalysed by enzymes.

The characteristics of heterogeneous catalysis are those outlined in Chapter 6A for catalysis in general. The reaction rate is proportional to the catalyst 'concentration' if this is regarded as the area of the phase boundary or solid surface, and the specific nature of heterogeneous catalysis is illustrated by the activity of enzymes, each of which will promote only certain types of change in certain molecules, and the decomposition of ethyl alcohol vapour at different solid surfaces

$$C_2H_5OH \begin{cases} Al_2O_3 \longrightarrow C_2H_4 + H_2O \\ 300°C \\ Cu \longrightarrow CH_3CHO + H_2 \end{cases}$$

The most thoroughly investigated heterogeneous processes are those involving gaseous reactants at a solid, usually metallic, surface. Since the rates of these reactions are expressed in terms of gas pressures, it is first necessary to relate gas pressure to the 'active reactant concentration', which in this case is *the amount of gas actually adsorbed at the solid surface.*

64

B. The Theory of Chemical Adsorption

Langmuir (1916) suggested that the adsorbed gas molecules are chemically linked to the atoms or molecules of the surface, the only difference between this and true compound formation being that the atoms of the surface are themselves bound together; thus the adsorption of an oxygen molecule at a tungsten surface may be visualised as

$$
\begin{array}{cccc}
 & \text{O---O} & & \\
 & \vdots \; \vdots & & \\
\text{---W---} & \text{W---} & \text{W---} & \text{W---} \\
| & | & | & | \\
\text{---W---} & \text{W---} & \text{W---} & \text{W---} \\
| & | & | & |
\end{array}
$$

This view is supported by the following experimental observations:—

(a) the heats of adsorption are of those associated with bond formation;

(b) a definite saturation limit appears to be reached when the surface is covered by a layer of gas one molecule thick;

(c) the catalytic effects of surfaces are specific.

The relation between the 'active concentration' and pressure of reacting gas is derived as follows:—

At temperature T and gas pressure p an equilibrium is established between the molecules adsorbed at the solid surface and those in the homogeneous phase such that

$$
\frac{\text{the rate of adsorption of}}{\text{molecules on to the surface}} = \frac{\text{the rate of desorption of}}{\text{molecules from the surface}}
$$

If under these conditions the fraction of unit surface area covered by adsorbed molecules is σ

the rate of adsorption $= ap(1 - \sigma)$

and the rate of desorption $= d\sigma$

where a and d are rate constants for adsorption and desorption; thus

$$ap(1 - \sigma) = d\sigma$$

or
$$\sigma = \frac{ap}{d + ap} \qquad (7.1)$$

Two limiting cases are recognised

(a) when σ is small, $\qquad 1 - \sigma \approx 1$

and $\qquad\qquad\qquad ap \approx d\sigma$

or $\qquad\qquad\qquad \sigma \approx ap/d \qquad (7.2)$

i.e. the amount of gas adsorbed is proportional to the gas pressure when the fraction of solid surface area covered is small;

(b) when $\sigma \approx 1$ $\qquad ap(1 - \sigma) = d$

or $\qquad\qquad\qquad (1 - \sigma) = d/ap$ $\qquad\qquad$ (7.3)

i.e. when the solid surface is almost completely covered the free surface area $(1 - \sigma)$ is inversely proportional to the gas pressure.

C. Kinetics of Heterogeneous Gas Reactions

According to the law of mass action the rate of a heterogeneous reaction will be proportional to σ which is a measure of the active reactant concentration. The expression for the observed rate, measured as the rate of change of reactant pressure in the homogeneous phase, will depend on the extent to which the reactants are adsorbed as well as on the molecularity. It will be assumed in the first place that the products are desorbed immediately they are formed.

Reactants weakly adsorbed

In this case the fraction of solid surface covered is small and is proportional to the gas pressure p (7.2). If the reaction is unimolecular the rate expression is

$$-\frac{\mathrm{d}p}{\mathrm{d}t} = k\sigma = \frac{kap}{d} = k'p$$

and first-order behaviour is exhibited. The decomposition of N_2O on a gold surface, PH_3 on a glass or porcelain surface, HCOOH on various surfaces and HI on a platinum surface are typical of the comparatively large number of reactions of this type.

If both reactants are weakly adsorbed the rate expression for a bimolecular heterogeneous process is

$$-\frac{\mathrm{d}p}{\mathrm{d}t} = k\sigma^2 = k\left(\frac{ap}{d}\right)^2 = k''p^2$$

and the reaction is second-order. The bromination of ethylene at the walls of a glass vessel

$$C_2H_4 + Br_2 \longrightarrow C_2H_4Br_2$$

is one of the few reactions which behave in this way.

In general if the reactants are weakly adsorbed, which is the case at higher surface temperatures, the order and molecularity of the reaction are identical.

Reactants moderately adsorbed

The use of the general equation (7.1) which is applicable in this case leads to the rate expression

$$-\frac{\mathrm{d}p}{\mathrm{d}t} = k\sigma = \frac{kap}{d + ap} \approx k'p^n$$

for a unimolecular process. The observed order n is less than unity and is temperature-dependent in so far as d increases with temperature; $n = 0\cdot6$ for the decomposition of SbH_3 on an antimony surface at 20°C. At higher temperatures where $d \gg ap$, the fraction of solid surface area covered is small and n approaches unity, as would be expected for weak adsorption.

Reactants strongly adsorbed

If the rate of desorption is much slower than the rate of adsorption, the surface is completely covered by reactant molecules regardless of the gas pressure, i.e. since $d \ll ap$ the rates of unimolecular and bimolecular processes are

$$-\frac{\mathrm{d}p}{\mathrm{d}t} = k\sigma = \frac{kap}{d + ap} \approx k \qquad (7.4)$$

$$-\frac{\mathrm{d}p}{\mathrm{d}t} = k'\sigma^2 = k'\left(\frac{ap}{d + ap}\right)^2 \approx k'$$

The rates are independent of gas pressure and the reactions are zero-order. It is moreover impossible to determine whether a single reactant is consumed in a unimolecular or a bimolecular process.

Integration of the zero-order rate expression (7.4) between the limits $p = p_0$ when $t = 0$ and $p = p_t$ after t sec leads to:

$$\left(p\right)_{p_0}^{p_t} = -k\left(t\right)_0^t$$

or $$p_0 - p_t = kt$$

Since $p_t = p_0/2$ when $t = t_{\frac{1}{2}}$ it follows that:

$$t_{\frac{1}{2}} = p_0/2k \qquad (7.5)$$

The rate of a zero-order reaction is independent of reactant pressure (7.4) and therefore independent of time, whilst the time of half-change is *directly proportional* to the initial gas pressure. This is illustrated by the data in Fig. 15 obtained for the decomposition of ammonia on a hot tungsten wire; the rate is equal to the slope of the curves which are almost identical for initial pressures of 50, 100 and 200 mm of ammonia and which deviate only slightly from

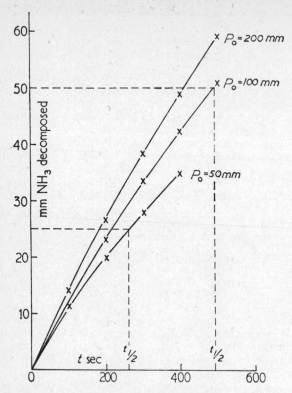

FIG. 15. Decomposition of NH_3 on tungsten at 856°C.
(*Hinshelwood and Burk, 1925*)

linearity associated with zero-order behaviour; moreover $t_{\frac{1}{4}} \approx 500$ seconds when $p_0 = 100$ mm and $t_{\frac{1}{4}} \approx 250$ sec when $p_0 = 50$ mm. Other examples of zero-order reactions are

the catalytic hydrogenation of unsaturated organic liquids
the decomposition of HI at a gold surface
enzymic reactions in solution (Section F).

Reactants adsorbed to different extents

If one reactant is adsorbed more strongly than the other the kinetics of a bimolecular heterogeneous process are quite complex.

However, in the limiting case where A is weakly adsorbed and B is strongly adsorbed the rate expression for the reaction

$$A + B \longrightarrow \text{products}$$

often has a simple form. The fraction of surface uncovered by B is (7.3)

$$(1 - \sigma_B) \approx \frac{d_B}{a_B p_B}$$

The fraction of this free surface covered by A is (7.2)

$$\sigma_A' = \frac{a_A p_A}{d_A}$$

so that the fraction of total surface covered by A is

$$\sigma_A = \sigma_A'(1 - \sigma_B) = \frac{a_A d_B p_A}{a_B d_A p_B}$$

Since the reaction is zero-order in B and first-order in A the rate expression becomes

$$-\frac{dp_A}{dt} = k\sigma_A = k\frac{a_A d_B p_A}{a_B d_A p_B} = k'\frac{p_A}{p_B}$$

i.e. the rate is inversely proportional to the pressure of the strongly adsorbed reactant B. This behaviour is exhibited by the reaction

$$2CO + O_2 = 2CO_2$$

at a quartz surface, the experimental rate expression

$$\frac{dp_{CO_2}}{dt} = k\frac{p_{O_2}}{p_{CO}}$$

indicating that whereas O_2 is weakly adsorbed, the quartz surface is virtually covered by an adsorbed layer of CO. In the same way the rate of the reaction

$$2H_2 + O_2 = 2H_2O$$

at a platinum surface which strongly adsorbs H_2, is inversely proportional to the pressure of this gas.

The fact that a reactant itself can 'inhibit' the rate of a reaction between two molecular species shows that both reactants must be adsorbed before they can react; if it were sufficient for an O_2 molecule in the gas phase to strike an adsorbed molecule of CO or H_2, the rates of these reactions should be independent of the pressures of CO and H_2, respectively, since the reactions are zero-order with respect to these reactants.

It has been assumed throughout this treatment that the products are desorbed as soon as they are formed; if they are not, the reaction is retarded as discussed in the following chapter.

D. The Order of Heterogeneous Reactions

The most useful criterion of the order n of a heterogeneous process is the dependence of the reaction half-life $t_{\frac{1}{2}}$ on initial reactant pressure p_0. As in the case of homogeneous reactions the general expression

$$t_{\frac{1}{2}} \propto 1/p_0^{n-1}$$

is applicable even to zero-order reactions where, since $n = 0$,

$$t_{\frac{1}{2}} \propto 1/p_0^{-1} = p_0$$

This is illustrated by the decomposition of ammonia on tungsten (Fig. 15).

The relationship between the experimentally observed order and the molecularity depends on the extent to which the reactants are adsorbed and may be summarised as follows:—

(a) if the reactants are weakly adsorbed, order and molecularity are identical;

(b) the reaction is zero-order if the reactants are strongly adsorbed and it is impossible to distinguish the molecularity in the case of a single reacting species;

(c) the order varies with temperature if the reactants are moderately adsorbed;

(d) since adsorption decreases as the temperature is raised, reactions of this type tend to exhibit zero-order behaviour at low temperatures whilst at higher temperatures the order observed approaches the molecularity.

E. The Temperature Dependence of Heterogeneous Reaction Rates

As in the case of homogeneous reactions, the experimental rate constant k for a heterogeneous process varies with temperature T according to the Arrhenius equation

$$\log_e k = \text{constant} - \frac{E}{RT} \qquad (7.6)$$

where E is the activation energy. Fig. 16 shows a plot of $\log_{10} t$ against $1/T$ for the unimolecular decomposition

$$2N_2O = 2N_2 + O_2$$

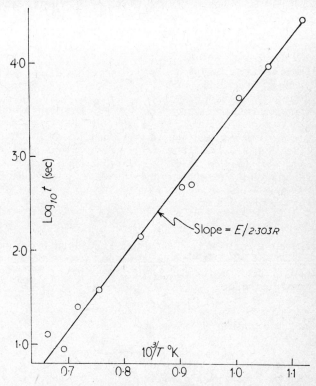

FIG. 16. Temperature dependence of N_2O decomposition on platinum. (*Hinshelwood and Prichard, 1925*)

at the surface of a platinum wire; t is the time required for 10% decomposition, and since $t \propto 1/k$, (7.6) becomes

$$\log_e t = \frac{E}{RT} + \text{constant}$$

The slope of the line in Fig. 16 is equal to $E/2\cdot303R$ whence

$$E = 32\cdot5 \text{ kcal/mole}$$

for this reaction which is considerably less than the value of $\sim$60 kcal/mole found for the uncatalysed homogeneous change.

The exponential temperature dependence of the rate of a heterogeneous reaction indicates that in addition to being adsorbed, the reactant molecules must also be activated, the sequence of events being:—

   (a) reactants are adsorbed;
   (b) adsorbed reactants are activated;
   (c) activated reactants $\longrightarrow$ products;
   (d) products are desorbed.

A heterogeneous process may therefore be regarded as a homogeneous reaction taking place on the two-dimensional surface.

In addition to increasing the rate at which adsorbed molecules react, however, *an increase in the surface temperature will also reduce the number of molecules adsorbed*, thereby lowering the active reactant concentrations. If the adsorption is weak the unimolecular rate expression is

$$-\frac{\mathrm{d}p}{\mathrm{d}t} = k\sigma = A\mathrm{e}^{-E/RT}\sigma \qquad (7.7)$$

where the rate constant k is related to the actual number of molecules adsorbed and E is the true activation energy. Since the rate constant for desorption is related to the heat of adsorption Q by an expression of the form

$$d = d_0\mathrm{e}^{-Q/RT}$$

where d_0 is a constant, (7.2) becomes

$$\sigma = \frac{ap}{d} = \frac{ap}{d_0}\mathrm{e}^{Q/RT}$$

and the expression (7.7) may be written

$$-\frac{\mathrm{d}p}{\mathrm{d}t} = A\mathrm{e}^{-E/RT}\frac{ap}{d_0}\mathrm{e}^{Q/RT} = A'\mathrm{e}^{-(E-Q)/RT}p$$

$$= k'p$$

The measured rate constant k' is expressed in terms of the total number of free and adsorbed reactant molecules present and is given by

$$k' = A'\mathrm{e}^{-(E-Q)/RT}$$

whence

$$\frac{\mathrm{d}\log_e k'}{\mathrm{d}T} = \frac{E-Q}{RT^2} = \frac{E'}{RT^2}$$

It follows therefore that

$$E' \qquad = \qquad E \qquad - \qquad Q$$

| the observed activation energy | the true energy of activation | the heat of adsorption |

If on the other hand the reactants are strongly adsorbed so that the surface is completely covered over the temperature range investigated, the active concentration of reactants is independent of temperature and the true and observed activation energies of these zero-order reactions are equal.

F. Enzyme Catalysis

Enzymes are proteins which catalyse reactions in living systems. Because their dimensions are in the colloidal range and because their kinetic behaviour is similar to that of heterogeneous processes, enzyme catalysis has been referred to as *microheterogeneous*.

The remarkable selectivity of enzymes, which forms a basis for their classification, is illustrated by the behaviour of the enzyme urease which catalyses the hydrolysis of urea

$$CO(NH_2)_2 + H_2O = CO_2 + 2NH_3$$

but which is ineffective in the hydrolysis of substituted ureas; and of β-glucosidase which promotes the hydrolysis of β-glucosides but which has no effect on the isomeric α-glucosides.

Kinetics of enzyme catalysis

Like other catalysts, an enzyme accelerates the approach to chemical equilibrium at a rate which is proportional to its concentration $[E]$. If the concentration of reactant or substrate S is sufficiently low the experimental rate expression for an enzyme-catalysed process is

$$- \frac{d[S]}{dt} = k_E[S][E] \tag{7.8}$$

At higher substrate concentrations, however, the reaction undergoes a transition from first-order to zero-order behaviour and the rate expression becomes

$$- \frac{d[S]}{dt} = k_E'[E] \tag{7.9}$$

This is illustrated by the curve in Fig. 17 for the inversion of sucrose catalysed by the enzyme invertase.

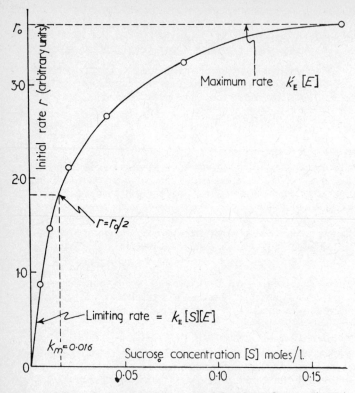

FIG. 17. Effect of substrate concentration [S] on rate of sucrose inversion catalysed by enzyme invertase (E).
(*Michaelis and Menten, 1913*)

Henri (1902) attributed this behaviour to the reversible formation of an enzyme–substrate complex ES which can decompose to products with simultaneous regeneration of the enzyme, thus

$$E + S \rightleftharpoons ES \xrightarrow{k} \text{products} + E$$

If $[E]_0$ and $[E]$ are the total and free enzyme concentrations, respectively, then

$$[E] = [E]_0 - [ES]$$

and the dissociation constant K_m of the complex ES is given by

$$K_m = \frac{[E][S]}{[ES]} = \frac{([E]_0 - [ES])[S]}{[ES]} \qquad (7.10)$$

Solving (7.10) for $[ES]$ leads to

$$[ES] = \frac{[E]_0[S]}{K_m + [S]}$$

whence the reaction rate r is

$$r = -\frac{d[S]}{dt} = k[ES] = \frac{k[E]_0[S]}{K_m + [S]} \qquad (7.11)$$

At low substrate concentrations when $[S] \ll K_m$, (7.11) reduces to

$$r = k[E]_0[S]/K_m \qquad (7.12)$$

which is identical with the experimental expression (7.8) with $k_E = k/K_m$; if on the other hand the substrate concentration is so large that virtually all the enzyme is complexed, (7.11) becomes

$$r_0 = k[E]_0 \qquad (7.13)$$

in agreement with (7.9). *An enzyme-catalysed process becomes zero-order therefore when the substrate concentration is sufficient to complex the enzyme completely, just as a heterogeneous gas reaction exhibits zero-order kinetics when the gas pressure is such that the solid surface is completely covered by adsorbed reactant molecules;* the actual substrate concentration or gas pressure necessary depends on the stability of the reactant–catalyst complex formed in each case.

If the enzyme concentration $[E]_0$ is eliminated from (7.11) and (7.13), the resulting expression

$$r = \frac{r_0[S]}{K_m + [S]} \quad \text{or} \quad K_m = [S]\left\{\frac{r_0}{r} - 1\right\} \qquad (7.14)$$

shows that K_m is equal to the concentration $[S]$ at which the reaction rate r is one-half the maximum zero-order rate r_0. Equation (7.14) is known as the Michaelis–Menten equation after its originators and the Michaelis constant K_m is a measure of the catalytic efficiency of an enzyme. From Fig. 17 it is seen that, when $r = r_0/2$,

$$[S] = K_m = 0.016 \text{ moles/lit}$$

Michaelis constants vary from 10^{-8} for very efficient enzymes to 1 for those which form relatively weak complexes with the substrate.

Temperature effects

So long as the enzyme itself is unaffected, the rate of an enzyme-catalysed process increases exponentially with temperature according to the Arrhenius expression. The experimental activation energies determined from the appropriate Arrhenius plots are much lower than those for the corresponding non-enzymic processes; thus the activation energy for the decomposition of H_2O_2 is reduced from 18 to 12 kcal/mole in the presence of colloidal platinum, but is only 5·5 kcal/mole when the enzyme liver-catalase is used as catalyst. Those physiological processes manifest in respiratory rhythms and beating of the heart have an overall activation energy of some 17 kcal/mole, whereas fireflies flash and crickets chirp at a frequency characterised by an activation energy of 12 kcal/mole. These remarkably low values are of course to be expected for biological processes to be recognised on our time scale at body temperature.

If the temperature is raised sufficiently, the enzyme is thermally inactivated by denaturation and the rate of enzymic reaction decreases abruptly. This behaviour is illustrated by Fig. 18, which shows an Arrhenius plot for the decomposition of hydrogen peroxide by catalase.·

Mechanism of enzyme catalysis

It is clear from the remarkably low activation energies of enzymic reactions that enzymes, like other catalysts, provide an alternative reaction path with a lower energy barrier than the uncatalysed process. The ability of the theory proposed by Henri to account for the kinetics of enzyme catalysis indicates that complex formation with the substrate does indeed take place and it is believed that the functional groups of the enzyme form active centres with which the substrate combines. If more than one active centre is involved in complex formation with the same substrate molecule, the selectivity of enzyme catalysis can be explained in terms of a critical configuration of active centres; this is the basis of the 'polyaffinity' theory. However, complex formation is not necessarily sufficient for reaction and other factors must play an important role. The difficulties involved in elucidating the mechanism of enzymic reactions are illustrated by the fact that at least twelve enzymes are involved in the comparatively simple fermentation of glucose

$$C_6H_{12}O_6 = 2C_2H_5OH + 2CO_2$$

in the presence of yeast cells.

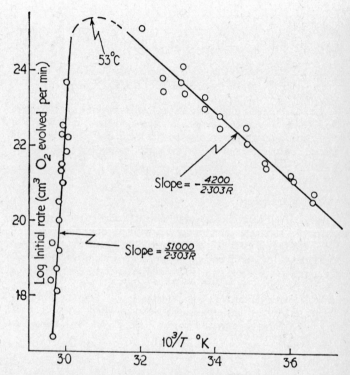

FIG. 18. Arrhenius plot for decomposition of H_2O_2 catalysed by enzyme catalase, $E = 4.2$ kcal/mole. Rate decreases above 53°C due to inactivation of catalase for which $E = 51 + 4.2$ kcal/mole.
(*Sizer, 1944*)

8

THE DEPENDENCE OF REACTION RATE ON PRODUCT CONCENTRATION

THE rate of chemical change is always proportional to the active concentration of reacting molecules; the measured rate will *also* depend on the concentration of products formed if (a) the products reform the reactants, or (b) the products catalyse or inhibit the reaction.

A. Opposing Reactions

In principle all reactions are reversible and the products re-form the reactants at a rate which increases with product concentration until equilibrium is established. If the equilibrium constant is much greater than unity, or if one of the products is removed from the reacting system as soon as it is formed, this reverse or opposing reaction is so slow that its rate may be neglected. In all other cases, however, the overall rate of chemical change will depend on the rate constants of forward and opposing reactions, which may be of the same or of different order, and both will appear in the rate expression.

Both reactions first-order

This is the simplest case in which the product B is an isomer of reactant A

$$A \underset{k_2}{\overset{k_1}{\rightleftharpoons}} B$$

The net rate of consumption of A is equal to the actual rate of consumption less the rate at which it is re-formed from B, i.e.

$$-\frac{d[A]}{dt} = k_1[A] - k_2[B] \tag{8.1}$$

Since $[B] = [A]_0 - [A]$ where $[A]_0$ is the initial concentration of A, (8.1) may be written:

$$- \frac{d[A]}{dt} = k_1[A] - k_2([A]_0 - [A]) \tag{8.2}$$

A solution of equation (8.2), which contains two unknown constants k_1 and k_2, requires the further relationship between them provided by the equilibrium condition

$$k_1[A]_e = k_2([A]_0 - [A]_e) \tag{8.3}$$

where $[A]_e$ is the equilibrium concentration of A. Eliminating k_2 from (8.2) and (8.3) provides

$$- \frac{d[A]}{dt} = k_1[A] - \frac{k_1[A]_e([A]_0 - [A])}{([A]_0 - [A]_e)}$$

$$= k_1[A]_0 \frac{([A] - [A]_e)}{([A]_0 - [A]_e)} \tag{8.4}$$

Rearranging (8.4) and integrating between the limits $[A]_0$ and $[A]$ at $t = 0$ and $t = t$ leads to

$$\int_{[A]_0}^{[A]} \frac{d[A]}{([A] - [A]_e)} = - \frac{k_1[A]_0}{([A]_0 - [A]_e)} \int_0^t dt$$

or since

$$\int \frac{d[A]}{([A] - [A]_e)} = -\log_e \frac{[A]_e}{([A] - [A]_e)} + \text{constant}$$

$$- \log_e \frac{[A]_e}{([A]_0 - [A]_e)} + \log_e \frac{[A]_e}{([A] - [A]_e)} = \log_e \left(\frac{[A]_0 - [A]_e}{[A] - [A]_e} \right)$$

$$= \frac{k_1[A]_0 t}{([A]_0 - [A]_e)}$$

i.e.

$$k_1 = \frac{[A]_0 - [A]_e}{t[A]_0} \log_e \left(\frac{[A]_0 - [A]_e}{[A] - [A]_e} \right) \tag{8.5}$$

and from (8.3)

$$k_2 = \frac{[A]_e}{t[A]_0} \log_e \left(\frac{[A]_0 - [A]_e}{[A] - [A]_e} \right) \tag{8.6}$$

The calculation of rate constants for opposing reactions therefore requires a knowledge of the equilibrium concentration $[A]_e$; if this is very small compared with the initial concentration $[A]_0$ the reaction goes virtually to completion, $k_2 \approx 0$, (8.6), and (8.5) becomes:

$$k_1 = \frac{1}{t} \log_e \frac{[A]_0}{[A]}$$

which is identical with expression (3.5) obtained for an unopposed first-order reaction.

An example of reversible isomerisation is provided by the rearrangement of styryl cyanide vapour

Both reactions second-order

$$A + B \underset{k_2}{\overset{k_1}{\rightleftharpoons}} C + D$$

If the concentrations of A and B are equal, the overall reaction rate is given by

$$-\frac{d[A]}{dt} = k_1[A]^2 - k_2([A]_0 - [A])^2 \tag{8.7}$$

and at equilibrium

$$k_1[A]_e^2 = k_2([A]_0 - [A]_e)^2 \tag{8.8}$$

Expression (8.7) can be integrated as before if k_2 is first substituted from (8.8).

This case is of particular interest since an example is the gas reaction

$$H_2 + I_2 \rightleftharpoons 2HI$$

investigated by Bodenstein; since it was not experimentally expedient to use equal concentrations of H_2 and I_2 the appropriate form of the integrated rate expression is particularly unwieldy, but by inserting the molecular concentrations found by analysis after certain reaction times, Bodenstein was able to obtain the values of the rate constants of both processes from a knowledge of the equilibrium conditions.

Forward and opposing reactions of different order

The overall rate expression for the reversible dissociation

$$A \underset{k_2}{\overset{k_1}{\rightleftharpoons}} B + C$$

is

$$-\frac{d[A]}{dt} = k_1[A] - k_2[B][C] \tag{8.9}$$

$$= k_1[A] - k_2([A]_0 - [A])^2$$

since $[B] = [C] = [A]_0 - [A]$

Use is made of the equilibrium condition

$$k_1[A]_e = k_2([A]_0 - [A]_e)^2$$

in order to integrate (8.9) as before. The hydrolysis of an ester in aqueous solution, e.g.

$$CH_3COOC_2H_5 + H_2O \rightleftharpoons CH_3COOH + C_2H_5OH$$

is kinetically of this type since with H_2O in excess the forward process is pseudo first-order.

The appropriate rate expression is obtained and treated in the same way for the association process

$$A + B \rightleftharpoons C$$

of which the isomerisation of an alkyl ammonium cyanate to the corresponding urea in aqueous solution

$$C_2H_5NH_3^+ + CNO^- \rightleftharpoons CO(NH_2)NHC_2H_5$$

is an example.

It is again emphasised that in order *to obtain the rate constants for all opposing reactions, a knowledge of the equilibrium conditions is necessary.*

B. The Products Catalyse the Reaction—Autocatalysis

If the process:

$$A \longrightarrow B + C$$

is catalysed by the product B, the initial rate

$$\frac{d[B]}{dt} = k[A]$$

increases as B accumulates and becomes

$$\frac{d[B]}{dt} = k[A] + k_B[A][B] \tag{8.10}$$

If $[B]_0$ is the concentration of catalyst either present initially or produced by the uncatalysed reaction before the catalysed process predominates, (8.10) may be written:

$$\frac{d[B]}{dt} = k_B([A]_0 - [B])([B]_0 + [B]) \qquad (8.11)$$

which on integration (by partial fractions) leads to

$$k_B = \frac{1}{t([A]_0 + [B]_0)} \log_e \left\{ \frac{[A]_0[B]}{[B_0][A]} \right\} \qquad (8.12)$$

Equation (8.12) may be rearranged to (8.13) which shows how $[B]$ increases with time

$$[B] = \frac{[A]_0 + [B]_0}{1 + ([A]_0/[B]_0)e^{-([A]_0 + [B]_0)k_B t}} \qquad (8.13)$$

This gives an S-shaped curve typical of autocatalytic reactions such as the acid-catalysed hydrolysis of esters and various biochemical processes.

Fig. 19 shows how the concentration of trypsin produced autocatalytically from trypsinogen increases with time; the solid curve drawn according to (8.13) with k_B obtained from (8.12) is characteristic of many growth processes.

C. The Products Inhibit the Reaction

An important example of this behaviour is provided by heterogeneous gas reactions in which the products are adsorbed at the solid surface and reduce the area available for reaction. In the simplest case

$$A \rightarrow B + C$$

where A is weakly adsorbed and one of the products B is strongly adsorbed the available surface area is reduced to (7.3)

$$(1 - \sigma_B) = \frac{d_B}{a_B p_B}$$

If σ_A is the fraction of this available surface utilised by reactant A, the fraction of the total surface area at which reaction takes place is $(1 - \sigma_B)\sigma_A$ and the rate expression becomes

$$-\frac{dp_A}{dt} = k(1 - \sigma_B)\sigma_A = \frac{kd_B}{a_B p_B} \frac{a_A p_A}{d_A} = k' \frac{p_A}{p_B}$$

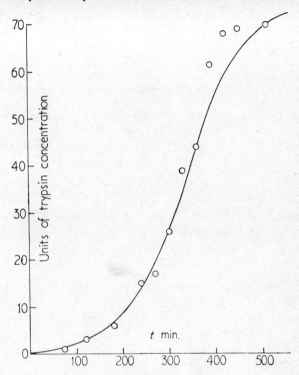

Fig. 19. Conversion of trypsinogen to trypsin which autocatalyses the reaction.

(*Kunitz and Northrop, 1936*)

i.e. the rate is inversely proportional to the pressure of product B. The hydrogen produced by the decomposition of ammonia on a platinum surface

$$2NH_3 = N_2 + 3H_2$$

is strongly adsorbed by the catalyst which accounts for the form of the observed rate expression

$$-\frac{dp_{NH_3}}{dt} = k\,\frac{p_{NH_3}}{p_{H_2}}$$

and the oxidation of sulphur dioxide at a platinum surface is similarly inhibited by the trioxide formed in the contact process for the manufacture of sulphuric acid

$$2SO_2 + O_2 = 2SO_3$$

Inhibition of enzymic reactions by reaction products frequently takes place. In this case the inhibitor competes with the substrate for the enzyme much as the adsorbed product competes with the reactant molecules for the catalytic surface in a heterogeneous gas reaction; thus the glucose formed from sucrose by the action of the enzyme invertase is a competitive inhibitor of this enzymic process

$$\text{sucrose} \xrightarrow{\text{invertase}} \text{glucose} + \text{fructose}$$

9

THE INFLUENCE OF RADIATION ON CHEMICAL REACTIONS

A. Electronically Excited Molecules

IN the reactions discussed so far, the activation energy is accumulated in the translational, rotational and vibrational degrees of freedom of the reacting molecules at a rate which increases with the temperature of the system; these processes are known as *thermal reactions*.

Like atoms, molecules also have electronic energy levels which can be excited by interaction with radiation; since the energy required to excite these levels is usually greater than the activation energies of thermal reactions, *electronically excited molecules can readily undergo chemical change*. The lowest excited electronic states of molecules are produced by the absorption of light in the visible or ultra-violet region of the spectrum and the study of their subsequent reaction forms the subject of *photochemistry*. Radiation of much higher energy, i.e. γ- and x-rays, as well as charged α- and β-particles, usually removes an electron completely from the molecules in its path; the study of reactions brought about by such ionising radiation is known as *radiation chemistry*.

Two consequences of electronic activation are

(a) the effect of temperature on reaction rate is much less pronounced than in the case of thermal reactions;

(b) the *overall* reaction brought about by radiation may be accompanied by an *increase* in free energy, in marked contrast to thermal processes which can only take place if they are spontaneous in the thermodynamic sense.

Electronic activation increases the free energy of the reactants such that the reactions of electronically excited molecules are accompanied by a free-energy decrease as required by the second law of thermodynamics. Examples of literally vital importance are the photochemical production of ozone in the upper atmosphere

$$3O_2 = 2O_3$$

and the photosynthetic reaction

$$mCO_2 + mH_2O = (CH_2O)_m + mO_2$$

which takes place in plants and which is an essential link in the carbon cycle. Both these processes are activated by light from the sun.

B. Photochemistry

Photochemistry is the study of reactions brought about by the absorption of light in the ultra-violet (1500—4000 Å) and visible (4000—7000 Å) regions of the spectrum, i.e. in the energy range of 40–200 kcal/mole; it is strictly only concerned with the primary processes undergone by electronically excited molecules, since secondary processes are usually of a thermal nature.

Light absorption

Clearly *a molecule can only react photochemically if it absorbs the light to which it is exposed;* this is the first law of photochemistry first recognised by Grotthus (1817) and Draper (1841). Furthermore *each molecule undergoing chemical change absorbs one quantum of light;* this is the law of photochemical equivalence (Einstein, 1905), often referred to as the second law of photochemistry.

Atoms will only absorb light of frequency ν if the energy $h\nu$ of the light quantum is exactly equal to the difference in energy between the unexcited or ground state E_1 and an excited electronic state E_2, i.e. the condition for absorption is

$$E_2 - E_1 = h\nu$$

or
$$\nu = (E_2 - E_1)/h \qquad (9.1)$$

The same condition holds for molecules except that since these have vibrational sub-levels which may be excited at the same time, the frequency of absorbed light is given by

$$(E_2 + V_2) - (E_1 + V_1) = h\nu$$

or
$$\nu = (E_2 - E_1)/h + (V_2 - V_1)/h \qquad (9.2)$$

V_2 and V_1 are the vibrational energies of the excited and ground state molecule, respectively. This is illustrated by reference to the potential energy curves for the ground and excited states of diatomic molecule X_2 shown in Fig. 20. At room temperature virtually all the molecules are in the lowest vibrational level of the ground state and transitions from this level to the different vibrational levels of the excited state, shown by the vertical arrows, will occur when the molecules are exposed to continuous radiation, that is light com-

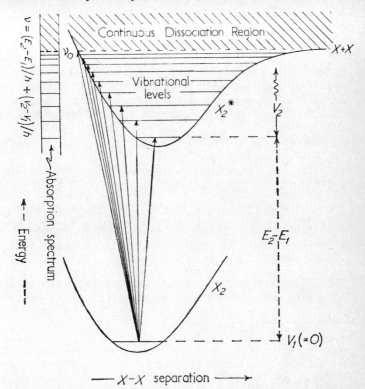

FIG. 20. Potential energy diagram for ground (X_2) and electronically excited (X_2^*) states of diatomic molecule X_2 showing transitions brought about by light absorption.

posed of all frequencies. The frequencies ν given by (9.2) are absorbed and will be missing from the light transmitted by the system as shown to the left of Fig. 20; this pattern of transmitted light is called an *absorption spectrum* and is characteristic of the absorbing molecule.

Primary photochemical processes

The absorption of a light quantum by the molecule X_2 may be represented by

(1) $$X_2 + h\nu \longrightarrow X_2^*$$

where X_2^* is an electronically excited molecule. X_2^* may then undergo one of the following primary processes, so called because they are the first things that can happen to an excited molecule.

(a) *Fluorescence emission;* this is the reverse of the absorption process (1) and may be written

(2) $$X_2^* \longrightarrow X_2 + h\nu$$

The absorbed quantum will always be re-emitted as fluorescence within $c. 10^{-8}$ sec of absorption unless one of the following processes takes place first.

(b) *Internal deactivation* represented by

(3) $$X_2^* \longrightarrow X_2$$

where the energy of the absorbed quantum is transformed into vibrational energy of the ground state.

(c) *Collisional quenching:*

(4) $$X_2^* + X_2 \longrightarrow X_2 + X_2$$

this is predominant at high gas pressures, where the collision frequency is high, and in solution.

(d) *Dissociation:*

(5) $$X_2^* \longrightarrow X + X$$

which takes place within a vibrational period if the molecule is excited in the continuous region of the spectrum, i.e. if ν is at least equal to ν_0 (Fig. 20).

Of these primary processes only the last is of photochemical interest since the others result in no chemical change; the atoms produced in process (5) are extremely reactive and undergo secondary processes with other molecules present to give the products of the overall reaction.

Polyatomic molecules have similar spectral characteristics and undergo the same primary processes. However, if the energy of the absorbed quantum is sufficiently great, the molecule does not dissociate into atoms only but into reactive fragments known as *free radicals*. Examples are

$$CH_3COCH_3 + h\nu \longrightarrow CH_3\!-\!\!- + \,-\!\!COCH_3$$
$$\text{acetone} \qquad\qquad \underset{\text{radical}}{\text{methyl}} \qquad \underset{\text{radical}}{\text{acetyl}}$$

$$CH_3CHO + h\nu \longrightarrow CH_3\!-\!\!- + \,-\!\!CHO$$
$$\text{acetaldehyde} \qquad \underset{\text{radical}}{\text{methyl}} \qquad \underset{\text{radical}}{\text{formyl}}$$

$$C_3H_7N_2C_3H_7 + h\nu \longrightarrow 2C_3H_7\!-\!\!- + \,N_2$$
$$\text{azopropane} \qquad\qquad \underset{\text{radicals}}{\text{propyl}}$$

Photochemical efficiency

A quantum of light can only excite one molecule. If the excited molecule then undergoes one of the primary processes (2), (3) or (4), this quantum is wasted from the photochemical standpoint. The efficiency of a photochemical reaction, i.e. the fraction of excited molecules undergoing chemical change, is known as the *quantum yield* γ more usually defined by

$$\gamma = \frac{\text{number of molecules reacting per second}}{\text{number of quanta absorbed per second}}$$

Consequently the rate of the photochemical reaction is

$$\text{number of molecules reacting per second} = \gamma I$$

where I, the number of quanta absorbed per second, is equal to the number of molecules excited per second. I has the same significance as the term $e^{-E/RT}$ in the rate expression for a thermal reaction in so far as both are a measure of the rate of activation; since I is proportional to the incident light intensity I_0, *the rate of a photochemical reaction is increased by an increase in light intensity just as the rate of a thermal reaction is increased by raising the temperature.*

The proportionality between the rate of photochemical change and the incident light intensity is utilised in the *photographic process*. A photographic plate consists of a suspension of small grains of silver bromide in a gelatin film. On exposure to light an electron is transferred from a bromine ion to a silver ion

$$Ag^+Br^- + h\nu \longrightarrow Ag + Br$$

so that after exposure some of the grains will contain traces of silver. When the plate is developed by immersion in a mildly reducing solution these grains will be largely reduced to silver whilst the others, which have received little or no light and in consequence contain no silver nuclei which promote reduction, are unaffected. The unreduced silver bromide is then removed by 'fixing' with sodium thiosulphate, and the density of the silver deposits left on the 'negative' is proportional to the intensity of light to which the original silver bromide grains were exposed.

Most photochemical changes are complex in that they involve subsequent reactions of the atoms or free radicals formed in the primary process; these secondary reactions are discussed in Chapter 10.

Photosensitisation

The quenching of an electronically excited molecule (process (4), p. 88) may be effected by a different molecular species Y_2 in which case the excited molecule Y_2^* can undergo chemical change although it does not absorb the light directly:

$$X_2^* + Y_2 \longrightarrow X_2 + Y_2^* \longrightarrow Y + Y$$

In this case the decomposition of Y_2 is *photosensitised* by X_2. This is extremely useful in bringing about the decomposition of molecules such as hydrogen and the paraffin hydrocarbons which do not absorb light readily transmitted by a glass or quartz reaction vessel. A commonly used photosensitiser is the mercury atom excited by light of wavelength 2537 Å (1 Å $= 10^{-8}$ cm) thus

$$Hg + h\nu \longrightarrow Hg^*$$
$$Hg^* + C_2H_6 \longrightarrow Hg + C_2H_5— + H$$

or
$$Hg^* + H_2 \longrightarrow Hg + 2H$$

The photosynthetic reaction is sensitised by chlorophyll molecules which absorb in the blue and red regions of the spectrum and transfer this energy to the reaction site.

C. Radiation Chemistry

Chemical change is produced by the exposure of molecules in the gaseous, liquid or solid state to α-, β-, γ- and X-rays, the energy of which may be millions of electron-volts (eV) compared with an equivalent energy of < 10 eV for ultra-violet or visible light. Consequently the effects of this high-energy radiation, which form the subject of radiation chemistry, differ from photochemical effects produced by the absorption of a light quantum in that

(a) an electron in an exposed molecule is not only excited but may be removed completely as a secondary electron in the process

$$M \rightsquigarrow M^+ + e^-$$

where $\rightsquigarrow$ denotes exposure to so-called *ionising radiation*;

(b) a single α-particle, β-particle, γ- or X-photon may ionise a large number of molecules directly along its *track* and indirectly along the branched track or *spur* of the secondary electrons; this leads to high local concentrations of excited molecules and ions which may react with each other prior to diffusion out of the spur;

(c) the absorption of ionising radiation is non-selective in that the major constituent of a two-component system (e.g. the solvent in a dilute solution) is preferentially ionised or excited.

The radiolysis efficiency (cf. the quantum yield of a photochemical reaction) is expressed as the number of molecules consumed or produced by the absorption of 100 eV of energy defined as the

G- value. For example the ^{60}Co γ-radiolysis of liquid ammonia or of pure cyclohexane may be represented by

$$(0.70)NH_3 \rightsquigarrow (0.22)N_2 + (0.81)H_2 + (0.13)N_2H_4$$

or $$cyclo\text{-}C_6H_{12} \rightsquigarrow (5.6)H_2 + (3.3)C_6H_{10} + (2.0)C_{12}H_{22}$$

where the G-values are given in parentheses.

In polar solvents one of the spur reaction products is the solvated electron e_{sol}^- which has a (solvent-dependent) absorption spectrum distinct from that of the solvent negative ion; this undergoes subsequent reaction with other spur products which also react with themselves and the solvent to generate the end radiolysis products. The complexity of the overall reaction is illustrated below by some processes undergone by probable spur products, in the pulsed β-radiolysis of pure water, for which rate constants have been measured:

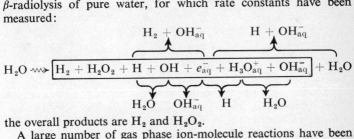

the overall products are H_2 and H_2O_2.

A large number of gas phase ion-molecule reactions have been studied in the ionisation chamber of a mass spectrometer at extremely low pressures. The primary ions, produced by electron bombardment and identified from the primary mass spectrum, react *in situ* with the added or parent gas at slightly higher pressures to produce secondary ions identified from the secondary mass spectrum. A number of these reactions are similar to those exhibited by uncharged free radicals (Chapter 10), e.g. H-atom transfer

$$H_2^+ + H_2 \longrightarrow H_3^+ + H$$
$$H_2^+ + C_2H_6 \longrightarrow H_3^+ + C_2H_5-$$
$$CH_4^+ + CH_4 \longrightarrow CH_5^+ + CH_3-$$
$$CH_4^+ + C_3H_6 \longrightarrow CH_5^+ + C_3H_5-$$

whereas others such as hydride ion (H^-) transfer, e.g.

$$C_3H_5^+ + C_5H_{12} \longrightarrow C_3H_6 + C_5H_{11}^+$$

and condensation reactions, e.g.

$$C_3H_5^+ + C_3H_6 \longrightarrow C_4H_7^+ + C_2H_4$$
$$CD_3^+ + CD_4 \longrightarrow C_2D_5^+ + D_2$$

have no counterpart in neutral free radical reactions.

10

ATOM AND FREE RADICAL REACTIONS

A. Primary and Secondary Processes

THE primary products of photochemical dissociation are usually atoms or free radicals (p. 88) which subsequently form the final reaction products in a series of secondary processes. Atoms and free radicals are also produced by thermal dissociation at sufficiently high temperatures, the activation energy of the primary thermal process

$$M \xrightarrow{kT} R_1 + R_3$$

e.g. $$CH_3CHO \rightarrow CH_3- + -CHO$$

being equal to the energy of the bond broken, $\sim$75 kcal/mole in the case of CH_3—CHO. The subsequent reactions of atoms and free radicals do not depend on the way in which they are formed and the final products of the thermal and photochemical decomposition of a compound in the gas phase are often identical.

Most important of these secondary processes are:

(a) H-atom abstraction from the reactant molecule M by the radical R_1 to generate a different radical R_2 and a product molecule P_1

$$R_1 + M \rightarrow P_1 + R_2$$

e.g. $$CH_3- + CH_3CHO \rightarrow CH_4 + CH_3CO-$$

(b) radical decomposition to a smaller radical R_1 and a second product molecule P_2

$$R_2 \rightarrow P_2 + R_1$$

e.g. $$CH_3CO- \rightarrow CO + CH_3-$$

(c) radical–radical combination to form a different product molecule P_3

$$R_1 + R_1 \rightarrow P_3$$

e.g. $$CH_3- + CH_3- \rightarrow C_2H_6$$

(d) atom or free radical addition to an unsaturated molecule which produces a larger radical

$$R_1 + M \longrightarrow R_4$$

e.g. $$CH_3- + C_2H_4 \longrightarrow C_3H_7-$$

This is the reverse of process (b) above and is important in free radical polymerisation reactions.

B. Chain Propagation

If the radical or atom R_2 produced in the secondary process

$$R_1 + M \longrightarrow P_1 + R_2$$

can regenerate the radical or atom R_1 either by decomposition or reaction with the substrate M, e.g.

$$R_2 \longrightarrow P_2 + R_1$$

a reaction chain is propagated in which the primary production of a single free radical R_1 can trigger the decomposition of a large number of reactant molecules M equal to the *chain length* of the overall reaction

$$M \longrightarrow P_1 + P_2$$

Radicals R_1 and R_2 are referred to as the *chain carriers*.

One of the earliest chain mechanisms was proposed by Nernst (1918) to explain the very fast reaction between gaseous hydrogen and chlorine exposed to light absorbed by chlorine in its dissociation continuum (Fig. 20). The products of the primary photochemical process

$$Cl_2 + h\nu \longrightarrow 2Cl$$

initiate the chain *propagating* steps

$$Cl + H_2 \longrightarrow HCl + H$$

$$H + Cl_2 \longrightarrow HCl + Cl$$

which are controlled by a chain *termination* step, in this case

$$Cl \longrightarrow \tfrac{1}{2}Cl_2 \text{ (at the vessel wall).}$$

The overall quantum yield of this reaction (p. 89) may be as high as 10^6, i.e. one million molecules of HCl are produced by each absorbed light quantum; since each quantum initiates two chains the average chain length is 500 000.

Probably the first free radical chain mechanism was suggested by

Taylor (1925) to account for the high quantum yield of the mercury photosensitised (p. 90) hydrogenation of ethylene in the gas phase. The reaction sequence

$$Hg + h\nu \longrightarrow Hg^*$$
$$Hg^* + H_2 \longrightarrow Hg + 2H$$
$$H + C_2H_4 \longrightarrow C_2H_5 —$$

$$C_2H_5 — + H_2 \longrightarrow C_2H_6 + H$$
$$H + C_2H_5 — \longrightarrow C_2H_6$$

includes the initiating, propagating and terminating steps essential to a chain reaction.

C. Characteristics of Chain Reactions

Reaction order

Despite their overall complexity the rates of radical chain reactions are often proportional to a simple (although not necessarily integral) power of reactant concentration or pressure. This is illustrated with reference to one Rice–Herzfeld reaction sequence in Table 10.1 where R_1 and R_2 are the chain carriers and the rate of chain initiation is equal to the rate of light absorption I for a photochemical process or to $k_i[M]$ for the thermal reaction.

For long chains where the propagating steps are almost entirely

TABLE 10.1. Example of a Chain Mechanism
(*Rice and Herzfeld*, 1934)

Chain step	Elementary process	Rate
Initiation	$M \xrightarrow{h\nu \text{ or } kT} R_1 + R_3$	I (or $k_i[M]$)
Propagation	$R_1 + M \longrightarrow P_1 + R_2$	$k_p[R_1][M]$
	$R_2 \longrightarrow P_2 + R_1$	$k_p'[R_2]$
Termination	$R_1 + R_1 \longrightarrow P_3$	$k_t[R_1]^2$

(M = decomposing molecule: P_1, P_2, P_3 denote stable product molecules; R_1, R_2 and R_3 are atoms or free radicals.)

responsible for the rate of reactant consumption this is given by (Table 10.1)

$$\frac{-d[M]}{dt} \simeq k_p[R_1][M] \tag{10.1}$$

where the carrier concentration $[R_1]$ cannot usually be measured. However, an application of the *stationary-state condition* defined by

$$\frac{d[R_1]}{dt} = \frac{d[R_2]}{dt} = 0$$

leads to the general relationship† for a controlled chain reaction:

rate of initiation = rate of termination

or $\qquad I \text{ (or } k_i[M]) = k_t[R_1]^2 \tag{10.2}$

The elimination of $[R_1]$ from equations (10.1) and (10.2) provides the rate expressions

$$\frac{-d[M]}{dt} \simeq k_p\left(\frac{I}{k_t}\right)^{\frac{1}{2}}[M] \tag{10.3}$$

for photochemical initiation (at low light intensities) and

$$\frac{-d[M]}{dt} \simeq k_p\left(\frac{k_i}{k_t}\right)^{\frac{1}{2}}[M]^{\frac{3}{2}} \tag{10.4}$$

for the thermal reaction in Table 10.1.

On the other hand if the chain terminating step in Table 10.1 is replaced by

$$R_2 + R_2 \longrightarrow P_4 \quad \text{rate} = k_t'[R_2]^2$$

equation (10.2) becomes

$$I \text{ (or } k_i[M]) = k_t'[R_2]^2 \tag{10.5}$$

and an application of the stationary-state condition provides the relationship

$$\frac{d[R_1]}{dt} \simeq k_p'[R_2] - k_p[R_1][M] = 0 \tag{10.6}$$

Accordingly from (10.1) and (10.6) the rate of reactant consumption is

$$\frac{-d[M]}{dt} \simeq k_p[R_1][M] = k_p'[R_2] \tag{10.7}$$

† This follows intuitively since if chains are formed and terminated at different rates the overall reaction would either accelerate to explosion or effectively stop.

and the elimination of $[R_2]$ from (10.5) and (10.7) provides the rate expressions

$$\frac{-d[M]}{dt} \simeq k_p'\left(\frac{I}{k_t'}\right)^{\frac{1}{2}} \tag{10.8}$$

and

$$\frac{-d[M]}{dt} \simeq k_p'\left(\frac{k_i}{k_t'}\right)^{\frac{1}{2}}[M]^{\frac{1}{2}} \tag{10.9}$$

for photochemical (10.8) and thermal (10.9) initiation. Thus the overall order of a chain reaction depends on the chain terminating step.

Under certain conditions equations (10.3) and (10.4) describe the photolysis and pyrolysis (thermal decomposition) of acetaldehyde vapour at high temperatures where dissociation of the acetyl radical CH_3CO- (R_2) is relatively rapid; this provides a strong indication that recombination of CH_3- (R_1) radicals is the important terminating step and accounts for the presence of small amounts of C_2H_6 (P_3) in the reaction products which are mainly CH_4 (P_1) and CO (P_2).

The industrially important polymerisation of unsaturated molecules can be initiated photochemically or thermally, usually in the presence of small amounts of initiators which readily produce free

TABLE 10.2. Radical Polymerisation of Monomer M

			Rate
Initiation	initiator $\xrightarrow{kT}$ R		k_i[initiator]
Propagation	$\begin{cases} R + M \\ RM- + M \\ \text{------} \\ RM_j- + M \end{cases}$	$\begin{array}{l} \longrightarrow RM- \\ \longrightarrow RM_2- \\ \text{------} \\ \longrightarrow RM_{j+1}- \end{array}$	$\begin{array}{l} k_p[R][M] \\ k_p'[RM-][M] \\ \text{------} \\ k_p''[RM_j-][M] \end{array}$
Termination	$\begin{cases} RM_j- + RM_k- \\ RM_m- + RM_n- \\ \text{------} \end{cases}$	$\begin{array}{l} \longrightarrow R(M)_{j+k}R \\ \longrightarrow R(M)_{m+n}R \end{array}$	$\begin{array}{l} k_t[RM_j-][RM_k-] \\ k_t'[RM_m-][RM_n-] \\ \text{------} \end{array}$

radicals (or ions) in the primary process; subsequent chain propagation involves addition of radicals to the monomer M to produce successively larger radicals which eventually recombine in the chain terminating step as illustrated in Table 10.2. If it is assumed that the propagation and termination rate constants are independent of radical size ($k_p = k_p' = k_p'' = \ldots$, $k_t = k_t' = \ldots$), the rate of monomer consumption is given by

$$\frac{-\mathrm{d}[M]}{\mathrm{d}t} \simeq k_p[M] \sum_{i=1}^{\infty} [RM_i-] \tag{10.10}$$

and the rate of initiation is equal to the sum of the rates of all possible termination processes, i.e.

$$k_i[\text{initiator}] = k_t \left(\sum_{i=1}^{\infty} [RM_i-] \right)^2 \tag{10.11}$$

Elimination of the total free radical concentration

$$\sum_{i=1}^{\infty} [RM_i-] = RM- + RM_2- + RM_3- + \ldots RM_j- + \ldots$$

from equations (10.10) and (10.11) yields the rate expression

$$\frac{-\mathrm{d}[M]}{\mathrm{d}t} = k_p \left(\frac{k_i}{k_t} \right)^{\frac{1}{2}} [\text{initiator}]^{\frac{1}{2}} [M] \tag{10.12}$$

which is first-order in monomer concentration as observed experimentally.

Temperature-dependence of reaction rate

The activation energy of a reaction (cf. Chapter 4) is given by

$$E = \frac{-R\mathrm{d} \ln k}{\mathrm{d}(1/T)}$$

where the experimental rate constant k relates the reaction rate to some power of reactant concentration (p. 14). Accordingly from the rate expression (10.3) the activation energy for the photochemical chain reaction in Table 10.1 is given by

$$E_{\text{photo}} = \frac{-R\mathrm{d} \ln (k_p I^{\frac{1}{2}}/k_t^{\frac{1}{2}})}{\mathrm{d}(1/T)}$$

$$= \frac{-R\mathrm{d} \ln k_p}{\mathrm{d}(1/T)} - \frac{R\mathrm{d} \ln I^{\frac{1}{2}}}{\mathrm{d}(1/T)} + \frac{R\mathrm{d} \ln k_t^{\frac{1}{2}}}{\mathrm{d}(1/T)}$$

$$= E_p + 0 - \frac{E_t}{2}$$

if I is assumed to be independent of temperature and k_p and $k_t^{\frac{1}{2}}$ are expressed in the Arrhenius form (p. 41)

$$k_p = A_p \exp(-E_p/RT)$$
$$k_t^{\frac{1}{2}} = A_t^{\frac{1}{2}} \exp(-E_t/2RT) \tag{4.6}$$

For a radical–radical addition terminating step the activation energy E_t is virtually zero in which case the overall activation energy is equal to that of the propagating step in which the reactant is consumed; e.g. for the photolysis of acetaldehyde

$$E_{\text{photo}} \simeq E_p \simeq 10 \text{ kcal/mole}$$

where E_p refers to the H-atom abstraction process

$$CH_3- + CH_3CHO \rightarrow CH_4 + CH_3CO-$$

On the other hand the activation energy for acetaldehyde pyrolysis at higher temperatures is 46 kcal/mole; from the appropriate rate expression (10.4) this is equal to

$$E_{\text{therm}} = \frac{-R \mathrm{d} \ln(k_p k_i^{\frac{1}{2}}/k_t^{\frac{1}{2}})}{\mathrm{d}(1/T)}$$

$$= E_p + \frac{E_i}{2} - \frac{E_t}{2}$$

$$= E_{\text{photo}} + \frac{E_i}{2}$$

whence $\qquad E_i = 2(E_{\text{therm}} - E_{\text{photo}}) = 72 \text{ kcal/mole}$

which is close to the C–C bond energy in CH_3CHO. This illustrates the general finding that the activation energy E_{therm} of a thermally-initiated chain reaction is appreciably less than that E_i of the initiating step but exceeds the activation energy for photochemical initiation of the same chain sequence (cf. p. 85).

The chain length

The chain length is defined as the number of reactant molecules consumed in a single chain initiated by a radical formed in the primary process, i.e. for long chains

$$\frac{\text{chain}}{\text{length}} = \frac{\text{propagation rate}}{\text{initiation rate} \times n} = \frac{-\mathrm{d}[M]/\mathrm{d}t}{nI \text{ (or } nk_i[M])} \tag{10.13}$$

Here n, the number of propagating radicals produced in the primary step, has a value of 2 for the photochemical reaction of H_2 and Cl_2 (see above) and is assumed to be unity for the decomposition of acetaldehyde (the relatively stable CHO– radical is believed to diffuse to the walls of the reaction vessel). From the definition of quantum yield γ (p. 89) it follows that the chain length for photochemical initiation is simply equal to γ/n which may be as high as 300 for acetaldehyde photolysis above 300°C.

A combination of equations (10.2) and (10.13) provides the alternative definition

$$\frac{\text{chain}}{\text{length}} = \frac{\text{rate of propagation}}{\text{termination rate} \times n}$$

which may be used to estimate chain lengths of thermal reactions from the relative amounts of products formed in propagating and terminating steps, i.e. the ratio P_1/P_3 for the reaction in Table 10.1 or the CO/C_2H_6 ratio in the pyrolysis products of acetaldehyde. On the other hand the average length of a polymerisation chain may be estimated from the average molecular weight of the final polymer if the chain terminating step is known, e.g.

$$\frac{\text{chain}}{\text{length}} = \frac{\text{average molecular weight of polymer}}{2 \times \text{molecular weight of monomer}}$$

$$= \text{degree of polymerisation}/2$$

if radical–radical addition is chain terminating as in Table 10.2.

D. The Inhibition of Chain Reactions

The addition of small amounts of propylene C_3H_6 or other inhibitors drastically reduces the pyrolysis rate of certain organic compounds in the gas phase due to the consumption of chain carriers R by the formation of more stable addition or abstraction products, e.g.

$$R + CH_3\text{—}CH{=}CH_2 \rightarrow P + \text{–}CH_2\text{—}CH = CH_2$$

The allyl radical which may be represented by the canonical structures

$$\text{–}CH_2\text{—}CH{=}CH_2 \longleftrightarrow CH_2{=}CH\text{—}CH_2\text{–}$$

is resonance stabilised and much less effective as a chain carrier than the radical consumed. However, the observation of a finite limiting reaction rate indicates either that the allyl radical eventually initiates new chains or that part of the uninhibited reaction proceeds by a non-chain molecular process; in either case propylene inhibition is often used as a criterion of radical chain propagation.

In certain reactions the identity of the chain carriers has been established by the addition of iodine which produces the corresponding iodide in the process

$$R + I_2 \longrightarrow RI + I$$

The overall effect of iodine however is often catalytic since the reduction in chain propagation rate may be offset by an increased rate of initiation in the sequence

$$I_2 \xrightarrow{kT} 2I$$
$$I + M \longrightarrow HI + R + P$$

promoted by the low bond energy of I_2.

Although the nature of their effect is not always understood, inhibitors are added to unsaturated organic liquids to prevent their thermal polymerisation during storage, and to solutions of hydrogen peroxide and other peroxides with low activation energies, which are susceptible to chain decomposition.

E. Chain Branching and Explosion

Chain propagating steps usually consume and generate equal numbers of chain carriers, the overall concentrations of which are time independent. If however the net result of a secondary reaction sequence is an increase in the number of chain carriers, this so-called *chain branching* leads to additional chain initiation, the reaction rate rapidly approaches infinity and explosion results, unless the branching is controlled by chain terminating steps.

This is illustrated with reference to the thermal reaction between hydrogen and oxygen in which the relevants steps are

$$\text{initiation} \longrightarrow -\text{OH}$$
$$-\text{OH} + H_2 \longrightarrow H_2O + H \qquad \text{propagation}$$
$$\left.\begin{array}{l} H + O_2 \longrightarrow -\text{OH} + O \\ O + H_2 \longrightarrow -\text{OH} + H \end{array}\right\} \text{ branching}$$

The overall result of the branching processes is

$$H + O_2 + H_2 \longrightarrow H + -\text{OH} + -\text{OH}$$

in which three chain carriers are produced at the expense of one corresponding to a *branching factor* of 3.

The gaseous oxidation of hydrocarbon mixtures initiated by a spark in the cylinder of an internal combustion engine also proceeds by a radical chain mechanism. Under the conditions of operation certain intermediate aldehydes and peroxides are relatively stable except at the high temperature of the end gas (the last portion to be burnt) when they decompose according to

$$M \longrightarrow R_1 + R_2$$

This process of *delayed branching* leads to a rapid increase in reaction rate and the resulting explosion, manifest as 'knock' or 'pinking' in the engine, may be alleviated by 'antiknock' additives such as $Pb(C_2H_5)_4$ which decompose to produce chain-inhibiting species.

PROBLEMS

1.1. From the standard free energy changes given calculate the maximum possible percentage conversion of reactants in the following reactions at 25°C and 1 atm; assume that the initial reactant concentrations are equal in each case.

	ΔG (cal/mole)
(a) $H_2 + Cl_2 = 2HCl$	$-45\ 540$
(b) $N_2 + O_2 = 2NO$	$41\ 400$
(c) $CH_3COOH + C_2H_5OH$	
$\qquad = CH_3COOC_2H_5 + H_2O$	780

(reactants 0·1 M in aqueous solution).

2.1. The saponification of aqueous methyl acetate is conveniently followed by estimating the hydroxyl ion concentration at certain intervals. The following data were found for a typical run at 25°C:—

time	0	3	5	7	10	15	min
$[OH^-]$	0·01	0·0074	0·0063	0·0055	0·0046	0·0036	moles/lit

Plot the concentration–time curve and determine: (a) the initial reaction rate; (b) the rate after 6 min.

2.2. The catalysed decomposition of a 0·1 M solution of H_2O_2 was followed by removing 25 ml samples at certain time intervals and estimating the peroxide concentration by titration against standard permanganate. From the data obtained:—

time	0	5	10	20	30	40	min
$KMnO_4$	30·7	24·7	19·9	13·1	8·2	5·7	ml

plot the concentration–time curve and determine: (a) the initial rate; (b) the rate after 10 min.

N.B. Since one of the products (O_2) is a gas the reaction proceeds to completion and the use of equation (2.7) is simplified; X in this case is the number of millilitres of $KMnO_4$ required.

3.1. The initial rate of the first-order decomposition of H_2O_2 at 40°C is $1·14 \times 10^{-5}$ moles lit^{-1} sec^{-1} (Fig. 1), the initial peroxide concentration being 0·156 moles/lit.

(a) Calculate the rate constant from the initial rate.

(b) Determine the reaction half-life from the rate constant and use Fig. 1 to check the value you obtain.

(c) Use equation (3.5) to find the times required for the reaction to reach 25% and 75% completion under these conditions.

3.2. Hinshelwood and Burk (1924) studied the thermal decomposition of nitrous oxide in the gas phase by recording the change in total pressure with time. The half-life was found to vary with initial pressure as follows at 757°C:—

initial pressure	52·5	139	290	360	mm
$t_{\frac{1}{2}}$	860	470	255	212	sec

Plot this data according to the equation (3.16) and find the reaction order.

3.3. The following data were found for the rate of inversion of sucrose under certain conditions.

time	0	30	90	130	180	min
[sucrose]	0·500	0·451	0·363	0·315	0·267	mole/lit

Find the order of reaction by the integration method and calculate the rate constant. Comment on the experimental order in relation to the stoichiometric equation

$$C_{12}H_{22}O_{11} + H_2O = C_6H_{12}O_6 + C_6H_{12}O_6$$

3.4. The pyrolysis of dimethyl ether vapour may be represented by the equation

$$CH_3OCH_3 = CH_4 + H_2 + CO$$

Hinshelwood and Askey (1927) found that the reaction is homogeneous under the conditions of measurement and obtained the following manometric data at 552°C:—

time	0	114	219	299	564	∞	sec
pressure	420	743	954	1054	1198	1258	mm

(a) Plot this data according to the integrated first-, second- and third-order rate expressions (3.5), (3.10) and (3.13), find which gives a consistent value for k and so deduce the reaction order. (Since pressure is proportional to concentration, it is not necessary to convert to concentration units to find the order, but it is necessary to calculate the partial pressure of unchanged ether from each total pressure reading using the stoichiometric equation.)

(b) What is the average value of k? (Unless the reaction is first-order, k from (a) will be in pressure units and must be converted.)

(c) Confirm the order by finding the ratio $t_{\frac{3}{4}}/t_{\frac{1}{4}}$ from the pressure–time curve and compare the rate constant obtained from $t_{\frac{1}{2}}$ with that found in part (b).

4.1. Calculate the energy of activation for the catalysed reaction
$$CH_3COOH + CH_3OH = CH_3COOCH_3 + H_2O$$
from the slope of the curve in Fig. 11C.

4.2. Determine the Arrhenius parameters A and E for the reaction
$$C_2H_5ONa + CH_3I = C_2H_5OCH_3 + NaI$$
from Fig. 11B, and write down the complete rate expression. What is the half-life of this reaction at 10°C when the reactant concentrations are both 0·05 moles/lit initially?

4.3. Crossley, Kienle and Benbrook (1940) obtained the following values for the rate constant of the reaction
$$C_6H_5N_2Cl = C_6H_5Cl + N_2$$

T	5	25	35	50	°C
k	$1·5 \times 10^{-6}$	$4·1 \times 10^{-5}$	$2·0 \times 10^{-4}$	$1·4 \times 10^{-3}$	sec^{-1}

Plot this data according to the Arrhenius equation (4.7) and determine the parameters A and E. How long would it take to collect 200 ml of N_2 from one litre of a 0·1 M solution of benzene-diazonium chloride decomposing at 10°C? (The dimensions of the rate constant provide a clue to the reaction order.)

4.4. From the data in Table 4.1 calculate the half-lives of

(a) the first-order gas reaction
$$C_2H_5Br = C_2H_4 + HBr \quad \text{at } 600°C$$

(b) the reaction
$$CH_3I + (C_2H_5)_3N = CH_3\overset{+}{N}(C_2H_5)_3 + I^-$$

in solution at 25°C when the initial reactant concentrations are 0·05 moles/lit.

4.5. From the data in Table 4.1 calculate the time required to produce 1·15 g of ethyl alcohol by saponification
$$CH_3COOC_2H_5 + OH^- = CH_3COO^- + C_2H_5OH$$
starting with 1 lit of 0·05 M ester solution

(a) at 25°C with $[OH^-]_0 = 0·05$ M;
(b) at 25°C with $[OH^-]_0 = 0·10$ M;
(c) at 35°C with $[OH^-]_0 = 0·03$ M.

4.6. From the data in Table 4.1 calculate the equilibrium constant for the system
$$2HI \rightleftharpoons H_2 + I_2 \quad \text{at } 350°C$$

5.1. If the collision diameter of HI is 4·0 Å calculate the collision number at 410°C. (1 Å = 10^{-8} cm; $R = 8·313 \times 10^7$ ergs mole^{-1} °C^{-1}; molecular weight of HI = 128.)

5.2. What fraction of the colliding molecules of HI have the necessary energy to react at 410°C if the activation energy is 44 500 cal/mole?

5.3. From the answers to the previous questions obtain the theoretical value for the rate constant: (a) in ml mole^{-1} sec^{-1}, (b) in lit mole^{-1} sec^{-1}; and calculate the P factor from the experimental value

$$k = 5 \times 10^{-4} \text{ lit mole}^{-1} \text{ sec}^{-1} \quad \text{at } 410°\text{C}.$$

6.1. Koelichen (1900) obtained the following rate constants at 25°C for the dissociation of aqueous diacetone alcohol in the presence of a hydroxyl ion catalyst:—

$[OH^-]$	0·0023	0·0094	0·018	moles/lit
k	$1·72 \times 10^{-5}$	$8·50 \times 10^{-5}$	$1·66 \times 10^{-4}$	sec^{-1}

$[OH^-]$	0·045	0·094	moles/lit
k	$4·32 \times 10^{-4}$	$8·3 \times 10^{-4}$	sec^{-1}

What is the mean value of the catalytic constant k_{OH^-} for this reaction at 25°C?

6.2. What would be the effect of increasing ionic strength on the rates of the following reactions?

(a) $CH_3COOCH_3 + OH^- = CH_3COO^- + CH_3OH$
(b) $S_2O_8^{--} + 2I^- = I_2 + 2SO_4^=$
(c) $2H^+ + 2Br^- + H_2O_2 = 2H_2O + Br_2$

6.3. Duboux and Favre (1936) found that the rate constant for the hydrolysis of diazoacetic ester

$$N_2CHCOOC_2H_5 + H_2O = HOCH_2COOC_2H_5 + N_2$$

increases with hydrogen ion concentration as follows:—

$[H^+]$	0·00046	0·00087	0·00158	0·00325	moles/lit
k	0·0168	0·0320	0·0578	0·1218	

Show graphically that the rate of this reaction is proportional to the catalyst concentration and determine the catalytic constant k_H for this reaction under these conditions.

6.4. King and Jacobs (1931) found that the rate constant for the reaction

$$S_2O_8 + I^- \longrightarrow SO_4^= + I_2$$

varies with ionic strength at 25°C as follows:—

$\mu =$	0·00245	0·00365	0·00445	0·00645	moles/lit
$k =$	1·05	1·12	1·16	1·18	lit mole^{-1} min^{-1}

$\mu =$	0·00845	0·01245	moles/lit
$k =$	1·26	1·39	lit mole^{-1} min^{-1}

Plot this data according to the Brönsted–Bjerrum equation (6.5) and find the charge on the persulphate ion. What is the limiting value of k at infinite dilution ($\mu = 0$)?

7.1. Fig. 15 shows the rate of decomposition of ammonia on a hot tungsten wire at different pressures; the reaction half-lives are given on p. 68. What is the average rate constant under these conditions?

7.2. The rate of decomposition of nitrous oxide at the surface of an electrically heated gold wire was measured by Hinshelwood and Prichard (1925) with the following results at 990°C:—

time	0	30	53	100	min
pressure	200	216	225	236	mm

(a) If the stoichiometric equation is

$$2N_2O = 2N_2 + O_2$$

calculate the final pressure and hence find $t_{\frac{1}{2}}$ from the pressure–time curve.

(b) At an initial pressure of 400 mm, $t_{\frac{1}{2}}$ was found to be 52 min at the same temperature. What is the order of this reaction?

(c) The rate constant increased with temperature as follows:—

T	834	938	990	°C
k	$4 \cdot 28 \times 10^{-5}$	$1 \cdot 23 \times 10^{-4}$	$2 \cdot 21 \times 10^{-4}$	

Find the activation energy.

8.1. Kistiakowsky and Smith (1934) obtained the following data for the reversible isomerisation of stilbene:

cis   *trans*

Temperature 280°C.

Time	0	1830	3816	7260	12006	∞	sec
% *cis*	100	88·1	79·3	70·0	48·5	17	

Calculate the rate constants of the forward and reverse processes. (It is unnecessary to convert percentages into concentrations when using equations (8.5) and (8.6) since the conversion factor cancels out; thus $[A]_0$ may be put equal to 100, $[A]_e = 17$, $[A] = \%$ *cis*. A value of k can be obtained for each of the four readings, or alternatively k can be determined from the slope of the appropriate plot.)

9.1. The photochemical oxidation of phosgene by oxygen in ultra-violet light may be represented by the equation:

$$2COCl_2 + O_2 = 2CO_2 + Cl_2$$

Rollefson and Montgomery (1933) found that the absorption of $4\cdot4 \times 10^{18}$ quanta resulted in the decomposition of $1\cdot31 \times 10^{-5}$ moles of phosgene at 2537 Å. What is the quantum yield of this reaction? (Avogadro's number is $6\cdot023 \times 10^{23}$.)

10.1. The term autoxidation refers to the slow reaction of most organic liquids or vapours with molecular oxygen at moderate temperatures represented by the chain sequence

$$
\begin{array}{ll}
M \rightarrow R- & \text{i.} \\
R- + O_2 \rightarrow RO_2- & \text{p.} \\
RO_2- + M \rightarrow RO_2H + R- & \text{p}'. \\
2RO_2- \rightarrow RO_2R + O_2 & \text{t.}
\end{array}
$$

Obtain a rate expression for this reaction assuming that the chains are long, and that process p is very much faster than p', and estimate the overall activation energy if $E_i = 50$ kcal/mole, $E_p' = 10$ kcal/mole and $E_t = 0$.

10.2. Use the stationary-state approximation to find a rate expression for the reaction in Table 10.1 if the chain terminating step is

$$R_1 + R_2 \rightarrow \text{products}$$

INDEX